Praise for
Ending the War on Artisan Cheese

"Catherine Donnelly serves up a compelling case for food regulation based on scientific evidence, not special interests, reminding us in the process of our duty to cherish and support small-scale, independent producers endangered by the encroachment of multinationals."

—**BRONWEN PERCIVAL**,
coauthor of *Reinventing the Wheel*

"Friends of raw milk cheese could not have a more knowledgeable, rational, and persuasive expert on their side than Catherine Donnelly. Detail by appalling detail, Dr. Donnelly lays bare the thin scientific support for many FDA regulations on food safety. Her exposé may enrage you, but that's the point. Informed and indignant consumers willing to fight for traditional foodways can win this war."

—**JANET FLETCHER**, publisher, Planet Cheese blog

"This bold and compelling book reveals the shocking truth behind why powerful industrial dairy producers are on a mission to impose a ban on all raw milk cheese, and how US regulators are blatantly misrepresenting credible scientific studies to intimidate and threaten the country's growing number of artisan cheesemakers. Make no mistake, the stakes are high. Unless the confrontation ends soon, artisan cheese with an authentic taste of place will disappear."

—**WILL STUDD**, host and executive producer, *Cheese Slices*

"One size does not fit all when it comes to food safety regulation. In this comprehensive, critical review of FDA policy and practice over the last three decades in regulating commercial cheesemaking, microbiologist Catherine Donnelly reveals how the twentieth-century industrial ethos that guides regulatory rule-making is dangerously out of step not only with the growing interest in producing and consuming artisanal foods, but also with the latest scientific evidence."

—**HEATHER PAXSON**, author of *The Life of Cheese*

"Dr. Donnelly's compelling scientific and historical arguments against arbitrary regulation and government overreach are must-reads for anyone who values raw milk artisan cheeses, one of our most delicious and safest traditional foods. Those seeking proof need go no further."

—**DAVID GIBBONS**, coauthor of *Mastering Cheese*; cheese columnist, *Wine Spectator*

"In a moment where alternative facts, the war on science, fake news, and the corporate consolidation of our food system are redefining our futures, this superbly written book is timely and necessary. Catherine Donnelly uses the experiences of cheesemakers, farmers, and scientists to explore the tension that exists among corporate behemoths, regulators, and cheesemakers working to build economic vitality and maintain traditions in rural communities. As a scientist, Dr. Donnelly exposes the political objectives that have dominated policy and enforcement at the FDA, expertly proposes alternative regulatory solutions, and defends the remarkable safety record cheese has enjoyed over millennia. Bravo!"

—**MATEO KEHLER**, cheesemaker, Jasper Hill Farm

"Dr. Donnelly has written a thorough and articulate book, one that explores how rules written far from the farm, by people with little to no farm experience, weaken food safety, and how current FDA policy does not jibe with current scientific understanding of microbiology. She argues that the government should not favor industrial producers over small farms, and better policy is not only possible, but already exists. She also explains how we can use our wallets and our pens to advocate for the return of common sense and science to food policy."

—**DAN STRONGIN**, former president, American Cheese Society; columnist, *Cheese Reporter*

"You'll find no better guide to the byzantine world of food safety regulation than the eminent Professor Donnelly, whose lucid prose illuminates the complex scientific and regulatory issues at the heart of this riveting story. At a time of sharp ideological division over the role of government, she explains that more regulation is not always better, but all regulation is not necessarily bureaucratic overreach. Going further she offers a common sense call for competence and science-based decisions. Donnelly is eloquent in reminding us that the world's food traditions are cultural treasures and unsparing in her condemnation of the shocking and even duplicitous behavior of the FDA."

—**FREDERIC C. RICH**, lawyer; environmentalist; author of *Getting to Green*

"Dr. Donnelly does the hard work of piecing together a definitive account of how our government has threatened artisanal cheese producers at home and abroad. I had previously read the published science and couldn't understand why the FDA would use it against US producers. This book aims to clarify their intentions. Consumers and elected officials now have a responsibility to take action to stop this overreach."

—**CARLOS YESCAS**, director, Oldways Cheese Coalition

"This amazing book is an insightful look into the story of American entrepreneurs who are under attack from faceless bureaucrats who are determined to homogenize a unique and centuries-old American treasure. The book sets out a cautionary tale for any person interested in the American dream—what is happening to the artisan cheese industry can easily happen to any other industry. This is a must-read for sophisticated people who care about the battle between regulation and prosperity."

—**GREGORY S. MCNEAL**, JD, PhD, professor, Pepperdine University School of Law; contributor, *Forbes*

"For over two decades, Dr. Catherine Donnelly has worked tirelessly to support and defend artisanal cheesemakers in the daunting arena of food safety assurance and regulation. *Ending the War on Artisan Cheese* presents a panoramic view of the scientific, regulatory, political, commercial, and legal landscape that now threatens the survival of traditional artisan cheese. Donnelly's epic account is a clarion call to work together toward a win-win regulatory model that supports large and small producers alike, those who employ cutting-edge technology and those who carry on centuries-old tradition. Let's move forward together!"

—**PAUL KINDSTEDT**, author of *Cheese and Culture*

ENDING *the* WAR *on* ARTISAN CHEESE

Also by Catherine W. Donnelly

The Oxford Companion to Cheese

Cheese and Microbes

ENDING *the* WAR *on* ARTISAN CHEESE

The Inside Story of Government Overreach and the Struggle to Save Traditional Raw Milk Cheesemakers

CATHERINE W. DONNELLY, PHD

Chelsea Green Publishing
White River Junction, Vermont
London, UK

Front cover photograph: Wheels of clothbound cheddar mature on their shelves within The Cellars at Jasper Hill. Image courtesy of Jasper Hill Farm, created by Lark Smotherton.

Project Manager: Alexander Bullett
Project Editor: Benjamin Watson
Copy Editor: Diane Durrett
Proofreader: Eliani Torres
Indexer: Lisa Himes
Designer: Melissa Jacobson
Page Composition: Abrah Griggs

Printed in Canada.
First printing November 2019.
10 9 8 7 6 5 4 3 2 1   19 20 21 22 23

Our Commitment to Green Publishing
Chelsea Green sees publishing as a tool for cultural change and ecological stewardship. We strive to align our book manufacturing practices with our editorial mission and to reduce the impact of our business enterprise in the environment. We print our books and catalogs on chlorine-free recycled paper, using vegetable-based inks whenever possible. This book may cost slightly more because it was printed on paper that contains recycled fiber, and we hope you'll agree that it's worth it. *Ending the War on Artisan Cheese* was printed on paper supplied by Marquis that is made of recycled materials and other controlled sources.

Library of Congress Cataloging-in-Publication Data
Names: Donnelly, Catherine W., author.
Title: Ending the war on artisan cheese : the inside story of government overreach and the struggle to save traditional raw milk cheesemakers / Catherine W. Donnelly.
Description: White River Junction, Vermont : Chelsea Green Publishing, 2019. | Includes bibliographical references and index.
Identifiers: LCCN 2019026522 (print) | LCCN 2019026523 (ebook) | ISBN 9781603587853 (paperback) | ISBN 9781603587860 (ebook)
Subjects: LCSH: Cheesemaking—United States. | Cheesemaking—Europe. | Raw milk cheese—United States. | Raw milk cheese—Europe. | Cheese industry—Law and legislation.
Classification: LCC SF271 .D684 2019 (print) | LCC SF271 (ebook) | DDC 637/.3—dc23
LC record available at https://lccn.loc.gov/2019026522
LC ebook record available at https://lccn.loc.gov/2019026523

Chelsea Green Publishing
85 North Main Street, Suite 120
White River Junction, VT 05001
(802) 295-6300
www.chelseagreen.com

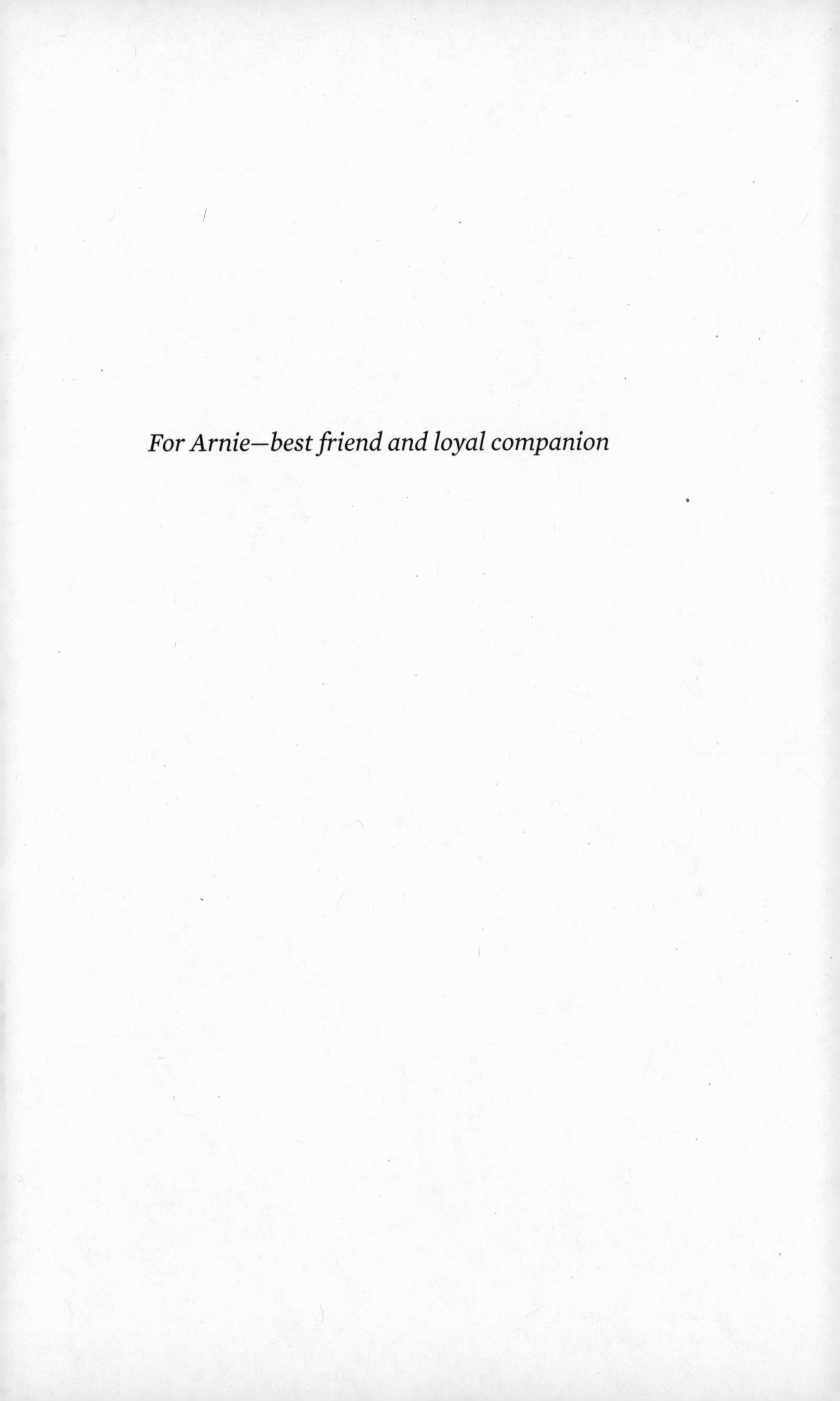

For Arnie—best friend and loyal companion

CONTENTS

PART THREE

What Do We Do Now?

ACKNOWLEDGMENTS

Taking on a project of this magnitude requires the assistance, support, and contributions of many individuals. First, I would like to thank my literary agent, Max Sinsheimer, for encouraging me to pursue this book. I would also like to thank Max for securing Chelsea Green as the publisher of this work. The entire Chelsea Green team—especially Margo Baldwin, Ben Watson, Alex Bullett, Diane Durrett, Melissa Jacobson, Eliani Torres, and Pati Stone—have provided wonderful support and encouragement throughout the publication process. Ben, in particular, has been a superb editor and incredible source of support and wisdom.

I would like to thank my colleagues in the artisan cheese world, who, through their everyday work, have informed so many parts of this writing. Thanks are extended to my good friends Mateo and Andy Kehler of the Cellars at Jasper Hill; Cathy Strange of Whole Foods; and my colleagues in the Cheese of Choice Coalition, who continue to fight to ensure that artisan cheeses, both domestic and imported, remain a part of the American culinary landscape.

I wish to thank Vermont's political leaders for their steadfast commitment to artisan cheese. Vermont's congressional delegation, particularly Congressman Peter Welch and members of his staff, including Mark Fowler, Ryan McLauren, George Twigg, Jake Oster, and Tricia Coates, provided lessons in leadership, collaboration, and reaching across the aisle when it was necessary. Tom Berry from Senator Leahy's office has been a tireless and effective advocate for promoting science to inform artisan cheese production and effective regulatory policies. Chuck Ross, in his role as Secretary of the Vermont Agency of Agriculture, Food and Markets promoted effective dialogue with federal regulatory officials that gave a voice at the table to Vermont's artisan cheese community.

I am grateful for wonderful scientific colleagues, including Paul Kindstedt, Sylvie Lortal, Giuseppe Licitra, and Stefania Carpino, who have greatly informed my knowledge of the science of cheese. And finally, I thank my friends and family who have supported me in so many ways during this writing.

PART ONE

Facts and Myths

CHAPTER 1

An Introduction to the Raw Milk Cheese Debate

By virtue of my scholarly interests and employment, I have had a front row seat in the raw milk cheese debate that has occurred both here in the United States and abroad over the last 35 years. I have personally witnessed the extraordinary renaissance in artisan cheesemaking taking place in the United States, particularly in my home state of Vermont, and also in England, Ireland, Quebec, Ontario, and throughout Europe.

The United States is the fastest-growing market for specialty cheese in the world. Beginning in the 1980s, when the pioneers of the American artisan cheese movement started making cheese, the industry has witnessed extraordinary growth in the number of producers and products, with corresponding improvements in quality and safety.[1] Over the last decade alone, the expansion has accelerated; since 2006 small-scale cheese companies, in some states, have doubled or even tripled in number. A bona fide cheese culture has blossomed here in the United States, and much of the demand is local. American cheeses now rival some of the very best produced worldwide. Artisan cheeses produced in the United States have become so sought after that companies based in Europe are buying US artisan cheese companies, seeing the potential for growth in artisan cheese consumption by US consumers. This should be a good news story for the US dairy industry, right?

Through a grant from the United States Department of Agriculture (USDA), I spent time in Ireland in 1990, learning from my colleague Dr. Charlie Daly about assistance the University College Cork (UCC) was providing to Irish artisan cheesemakers. We modeled our educational programs at the Vermont Institute for Artisan Cheese at the University of Vermont using the UCC model, and over a period of 13 years provided education and training on cheese safety to over 2,000 US artisan cheesemakers. We shared our educational resources with broader audiences, and for a time engaged in collaboration with the Innovation Center for US Dairy to promote artisan cheese safety. Things began to change on the regulatory front, however, and the very information and advocacy we were providing began to place the artisan cheese community under uncomfortable regulatory scrutiny. It became apparent that the regulatory approach of the US Food and Drug Administration (FDA) toward artisan cheesemakers was not based on sound science but rather on intimidation and fear, an arbitrary and capricious approach that will be explained in the chapters that follow. It is a fascinating and unfortunate history, one that should give us pause as we consider who makes decisions about the foods we eat, and the use of food safety as a guise for food choice. *Forbes* magazine said it best when, in 2014, it ran a story titled "FDA May Destroy American Artisan Cheese Industry."[2] This story has received over 216,000 views to date. For many cheese lovers, this was their first introduction to the fight between the FDA and the artisan cheese industry. However, as I will document, this has been a war long in the making. This book will provide an in-depth analysis of why *Forbes* was correct in its assessment.

Much of the growth of the US artisan cheese industry has been championed by the American Cheese Society (ACS). I first attended an ACS Annual Meeting in August 1993 at the invitation of its founder, Cornell Professor Frank Kosikowski. The meeting was conveniently held at the iconic Coach Barn on the grounds of Shelburne Farms, a bucolic estate that sits on a landscape designed by Frederick Law Olmsted and was the Vermont summer residence of Lila Vanderbilt Webb. Shelburne Farms produces a farmhouse Cheddar cheese from the milk of Brown Swiss cows that graze on the pastoral landscape of the farm. I participated as a cheese judge, and as I recall, there were 50 cheeses entered in the cheese judging and competition. I must admit that 10 of the 50 cheeses appeared defective and I refused to judge them. Fast-forward to 2018 when, at its annual meeting held in Pittsburgh, 1,954 cheeses were entered in the ACS competition. Two Vermont cheeses, Harbison and Calderwood, received first and second place Best in Show honors, an unprecedented achievement for the Cellars at Jasper Hill, a company that first debuted its cheese at ACS in 2003.

Today, the ACS Judging and Competition, the largest of its kind for judging cheeses made in the United States, requires about 30 judges to spend 3 full days in evaluation of the almost 2,000 cheeses entered in the competition. Best in Show winners often win global praise, going on to earn World Cheese Awards. The Festival of Cheeses, during which all entries from the ACS Judging and Competition are displayed and available for tasting by the general public, pays tribute to the remarkable efforts of US artisan and farmstead cheesemakers.

American artisan cheese has become mainstream, providing big business for retailers such as Whole Foods, Costco, Wegmans, Murray's Cheese (now owned by Kroger), and others. Despite the success enjoyed by US artisan cheesemakers and the meteoric rise of artisan cheese production, the American artisan cheese industry faces an existential threat: regulatory overreach. Regulations that govern cheesemakers in the United States apply equally to large industrial cheesemakers and small farmstead producers who use traditional practices in their cheese manufacture. Over the past 30 years, the FDA has pushed for a mandatory requirement for use of pasteurized milk in cheesemaking, claiming a public health risk for raw milk cheese. Under such a requirement, all raw milk cheeses would be banned.

That would be a travesty for cheese lovers. Chemical and microbiological analysis shows that raw milk cheeses are more complex and more flavorful than their pasteurized counterparts. The now classic 2002 *New Yorker* article "Raw Faith" by Burkhard Bilger provides an apt description of the difference, as written from his perspective during a visit with *maître fromager* Max McCalman at New York City's famed Picholine restaurant. From Picholine's high-tech cheese cave, Bilger writes:

> McCalman reached over and cut wedges from two Reblochon-style cheeses, one of pasteurized milk, the other of raw. We had done a few of these comparisons already, with the pasteurized invariably tasting milder, gummier, and less complex. But this time the difference was more elemental. The pasteurized version wasn't bad, with its musty orange rind and rich ivory pâte. But the raw-milk Reblochon seemed to bypass the taste buds and

> tap directly into the brain, its sweet, nutty, earthy notes rising and expanding from register to register, echoing in the upper palate as though in a sound chamber. I thought of something one of the founders of the Cheese of Choice Coalition had said when I asked her what difference raw milk could possibly make: "One is a cheese; the other is an aria by Maria Callas."

And as McCalman expressed to Bilger: "To eat a cheese like this was to participate in the preservation of a dying culture."[3]

Why the difference in the taste of cheeses made from pasteurized versus raw milk? Put simply, pasteurization kills off much of the natural milk flora responsible for flavor development. Traditional cheeses are fermented foods that are dependent upon the many microorganisms that contribute to the texture, flavor, veins, rinds, wrinkles, smells, and terroir of cheese. Additionally, in an artisan/farmstead system, milk that is produced from animals fed on pasture and/or dry hay has a high microbiological quality, which is essential for making the best cheese. In an industrial model, animals are confined indoors and often fed silage or other forage mixture. Silage is the source of many microorganisms that are detrimental to milk quality and safety. Milk is collected from many farms, transported in tanker trucks, and stored in refrigerated silos for many hours or even days prior to use. Spoilage organisms that grow under refrigeration produce heat-resistant enzymes that begin to attack milk proteins and fats, producing compounds that negatively alter the taste (the five senses perceived by the tongue: sweet, sour, salt, bitter, and umami) and flavor (the smell, texture, and expectation) of the milk. Government regulation requires pasteurization for milk produced under this industrial model to ensure both quality (destruction of spoilage organisms) and safety (destruction of pathogens). In contrast, in an artisan production model, controlling feed quality, closely monitoring animal health and hygiene, and transforming milk quickly into cheese bypasses these destructive steps and results in a cheese that reflects the superb attributes of the starting milk.

The debate over the safety of raw milk cheese has fiercely divided American cheesemakers and government regulators, and outraged cheese lovers. Currently, certain cheeses can be legally manufactured from raw milk only if they are aged for 60 days or longer. Aged cheeses have enjoyed a long and well-documented record of food safety, and the FDA's attempts to mandate pasteurization of all milk intended for cheesemaking comes despite scientific evidence supporting this record of safety. The FDA's activity has escalated recently, with the establishment of stringent microbiological criteria that only cheeses made from pasteurized milk can easily meet. When artisan and farmstead cheesemakers voiced their concern to the FDA through their congressional representatives in late 2015, the FDA temporarily backed down, but these cheesemakers remain extremely fearful for their regulatory future.

That is because, besides making our cheese more bland and the range of available cheese styles much narrower, a ban on raw milk cheesemaking would economically devastate nonindustrial cheesemaking in the United States. This is particularly true in states like Vermont, New York, California, Washington, and Wisconsin, where artisan cheese producers use raw milk in the production of aged Cheddar and other value-added cheeses. The artisan cheese renaissance is creating precious opportunities for small-scale dairy farmers and cheesemakers, who are often in rural economies facing challenging times. Jasper Hill Farm is one example of an artisan creamery that is flourishing in the Northeast Kingdom of Vermont, a region where until recently many barns hadn't seen cows in 40 or 50 years because the low price of milk bankrupted farmers. If artisan and farmstead cheesemakers are not able to differentiate their products from industrial, pasteurized processed cheeses, things will head south for them once more, and consumers will be deprived of the delicious cheeses that artisans produce. Mateo Kehler, the cofounder of Jasper Hill, put his frustration bluntly in the foreword to *The Oxford Companion to Cheese*: "I used to believe that the greatest threat to our business was a microbiological threat, but have learned the microbiological risk can be managed. I now believe the biggest risk to the cheeses that are the foundation of our business is a regulatory risk."

The weight of scientific evidence does not land in the FDA's favor. Most outbreaks occur as a result of post-process contamination, which is when cheese is recontaminated at some point along the production and distribution chain, negating the safety impact of pasteurizing the milk. Additionally, studies of several cheese styles have documented the ability of a host of bacterial pathogens—including *Listeria*, *Salmonella*, *E. coli*, and *Staphylococcus aureus*—to survive the current 60-day aging period, and yet outbreaks of human illness linked to raw milk cheese consumption are rarely reported. This remains true despite the tremendous growth in sales and consumption of artisan and traditional cheeses worldwide, which has exponentially increased consumer exposure to raw milk cheeses with limited adverse consequences.

This is not just an issue for American cheesemakers: A raw milk cheese ban could eliminate the ability of cheese retailers such as Whole Foods to import traditional cheeses from Europe and the rest of the globe because their PDO (Protected Designation of Origin), AOP (Appellation d'Origine Protégée), and former AOC (Appellation d'Origine Contrôlée) statuses often require these cheeses to be manufactured from raw milk. Beloved European cheese varieties such as Parmigiano Reggiano, Grana Padano, Gruyère, Comté, Emmental, and Roquefort would all be affected by proposed FDA regulations, and the importation of traditional cheeses such as Tomme de Savoie, Morbier, Abbaye de Belloc, St. Marcellin, Montbriac, Tomme de Bordeaux, and St. Nectaire *fermier* is already being restricted, or altogether eliminated. The issue has the potential to affect global trade and to further divide the United States from the rest of the world when it comes to views on food production and food safety.

Nor is America alone in imposing regulations that imperil the artisan cheese industry. Europe's beloved small producers are going belly-up or being bought out at unprecedented rates, in large measure because industrial dairy giants have lobbied regulators to be allowed to place PDO, AOP, or AOC labels on their products—designations that were intended for traditional products. In France in 2007–2008, Lactalis and Isigny Sainte-Mère, which together make more than 80 percent of traditional raw milk Camembert in Normandy, announced that they would begin using thermized and industrially treated milk in their Camembert production. Camembert's AOC standard requires the use of raw milk, but the dairy giants applied to have the standard rewritten so that they could keep the valuable AOC label while dramatically reducing the cost of their inputs—much cheaper milk, and fewer inspections now that they weren't using *lait cru* (raw milk). Lactalis and Isigny Sainte-Mère cited a public health concern in seeking the change: a 2005 case in which six children became ill after eating Camembert. What became known as the Camembert wars ended with the brands devastated by public perception that they had sold out their nation's gastronomic treasure for a cheap knockoff, imperiling the little producers who couldn't compete on price with mass-produced Camembert. Real Camembert activists marshaled the public opinion forcefully against the dairy giants, and Lactalis and Isigny Sainte-Mère dropped their pursuit of the AOC label. Instead, they were permitted to label their product *Fabriqué en Normandie*, or Made in Normandy, with pasteurized milk from any breed of cow permitted for use in Camembert manufacture.

The seeming victory of David over Goliath to preserve the AOC standard may have been short-lived, however. In February 2018, as reported by *Forbes* in the article "Why Your Genuine French Camembert Cheese Is in Danger," France's Institute of Origin and Quality announced that beginning in 2021 there will be a single AOP Normandy Camembert that includes cheese made with pasteurized milk, provided that at least 30 percent of the milk is obtained from Normande cows that graze in the region.[4] To many, this signaled the death of Normandy Camembert. An article published by Bloomberg titled "Why Camembert, One of the World's Great Cheeses, Might Soon Be Extinct" described what is at stake for Camembert du Normandie AOC producers. As the author, Larissa Zimberoff, writes:

> A PDO Camembert de Normandie must be made with unfiltered raw milk, with a fat content of at least 38 per cent that comes from cows from France's northern Normandy province, fed under strict conditions—grass and hay from local pastures. The milk must be hand-ladled in four or more layers into specific moulds. Milk is transported no farther than the distance that cows can slowly dawdle in search of a fresh blade of grass.[5]

Corroler and others conducted an ecological study to determine the effect of the geographic origin of specific bacterial strains on the manufacture and ripening of a traditional Camembert du Normandy cheese.[6] The consistent and specific presence of wild-type strains of the bacterium *Lactococcus lactis* isolated from raw milk produced within the AOC Camembert region confirmed the dairy significance of the Camembert region. As the authors stated: "It is well known that traditional cheeses made with raw milk ripen faster and develop a more intense flavor than cheeses made with pasteurized or microfiltered milk." Understanding the biodiversity of the microbial population associated with artisan cheese affords a look into the uniqueness that artisan production contributes to a biodiverse microflora in cheese, which in turn imparts unique sensory attributes.

My dear friend and colleague Dr. Sylvie Lortal, research director at the French National Institute for Agricultural Research (INRA), describes AOP as "gastronomic precious heritage." Meanwhile, centuries-old cheese styles like Fourme d'Ambert and Cantal are at risk of being lost because new health ordinances make them unaffordable or illegal to produce. In an impassioned speech to France's Institute of Sciences and Arts in acceptance of the François Rabelais prize for his organic farming efforts, the Prince of Wales, HRH Prince Charles, decried the "'bacteriological correctness' of European regulators." He asked, "In a microbe-free, progressive, and genetically engineered future, what hope is there for old-fashioned Fourme D'Ambert, the malformed Gruyère de Comté, or the odorous Pont L'Eveque?" He went on to eloquently state, "The distinctiveness of local cuisine is one of the most important ways we identify with the places and regions we love," adding that "a very important part of the whole magnificent edifice of European civilization rests on the inherited genius and craftsmanship of the people who make such distinguished concoctions" referring, of course, to those magnificent products we call cheese.[7]

Milk pasteurization is not the only issue at play in the global assault on traditional cheesemaking. In addition to the FDA's attempts to ban the use of wooden boards in cheese aging (many traditional cheeses made from both raw and pasteurized milk use aging on wooden boards to control cheese moisture levels and to allow development of flavor and character), the FDA has also tried to ban the use of ash in cheesemaking, a traditional practice that selects for important microbial groups during ripening of some bloomy rind cheeses made from pasteurized milk. And the FDA and other governments have established stringent *E. coli* standards that many artisan cheeses, regardless of whether produced from raw or pasteurized milk, simply cannot meet. Additional challenges artisan cheesemakers face include the FDA's *Listeria* swabbing assignments, the Food Safety Modernization Act (FSMA), and the harsh criminalization of outbreaks that are traced to their facilities.

Writing about raw milk cheese presents an opportunity to explore centuries-old food traditions, and to sound a warning to consumers that without vigilance these traditions may be eliminated for reasons that are simply not scientifically justified. *Ending the War on Artisan Cheese* ultimately provides focus on the politics of this issue, through an exploration of ways in which the public is rarely, if ever, consulted about the foods that they are allowed to consume. Simply put, it is highly unlikely that food safety concerns alone are at the heart of this complex issue. Governments taking part in the punishing regulatory activities have large industrialized dairy processors to please (Kraft, Fonterra, Lactalis, and the like). Most critical, however, is the unprecedented new regulatory authority given to the FDA under the FSMA to ensure food safety. With this new authority, there must be a system of checks and balances to ensure that this power is not abused, and that regulatory decision-making is based on sound science. The raw milk cheese debate provides a convenient and cautionary lens through which we can examine the FDA's functioning. As evidenced throughout the chapters that follow, I maintain that the FDA is not the right agency to oversee food safety. I hope this book will energize cheese lovers and, through their advocacy, lead to a new system of regulation that is grounded in science and that simultaneously supports food safety and promotes our precious rural working landscapes here in the United States and across the globe.

CHAPTER 2

The Historical Context for Raw Milk Cheesemaking

The following poignant obituary appeared in London's *Daily Telegraph* in May 2008:

Obituary of Lucy Appleby

Lucy Appleby, who died on April 24 aged 88, was one of the most accomplished cheese-makers of her generation and took a bold and ultimately successful stand against attempts to have unpasteurised cheese-making banned in Britain; with her late husband, Lance, she fought a battle against the standardisation of artisan cheese, in particular her own Appleby's Cheshire.

One of eight children, she was born Florence Lucy Walley on February 1 1920 at Lighteach Farm, Whitchurch, in Shropshire.

She attended Whitchurch High School and later went on to Reasheath Agricultural College, Nantwich, to learn cheese-making under the matriarchal figure of a Miss Bennion, an inspirational teacher who greatly influenced her belief in the goodness of unpasteurised and unwaxed Cheshire cheese.

Soon after leaving college, and by now an accomplished cheese-maker herself, Lucy met a local farmer, Lancelot Appleby, whom she married in 1940.

Two years later the couple moved to Hawkstone Abbey Farm, Shropshire, where for much of the next decade she was busy bringing up her family of seven children.

By 1952 she was ready to resume cheese-making, which she did using the unpasteurised milk from her herd of Friesian cows, a traditional recipe and calico binding, just as wax binding was coming into vogue.

The Applebys insisted on using calico, complaining that wax did not allow the cheese to breathe, thereby adversely affecting the flavour. The cloth is still used to bind their cheese today.

In the early years the cheese was sold through the Milk Marketing Board (MMB), and it was not until the early 1980s that it was marketed using the family name.

Most cheese-makers saw their salvation in dealing with the big supermarkets, and made industrial quantities using industrial methods. Lucy Appleby refused to follow suit, and stuck to producing her cheeses in the artisan style. In the early days this was often to the detriment of the business.

By 1982 Lucy Appleby had severed her links with the MMB, and found herself with a large supply of unsold cheese. At this point she approached Randolph Hodgson, the proprietor of Neal's Yard Dairy, a business specialising in selling artisan cheeses.

Hodgson recalled a Land Rover pulling into his premises: "Out of the back came a 40lb wheel of Appleby Cheshire cheese. I had never tasted anything like it before in my life. All the Cheshire cheeses I had been selling just paled away; this was a totally different cheese."

Hodgson's selling and marketing skills were to prove a turning-point. And in conjunction with Hodgson, Lucy Appleby formed the Specialist Cheese Makers' Association to lobby for the preservation of unpasteurised cheese.

Lucy Appleby was generous with her knowledge, but maintained a healthy sense of self-criticism. Often she would show visitors through the cheese room at the farm muttering: "This is a good batch, but that one is not. The colouration is not good in that one," and so on.

Ever present in her farmhouse kitchen was a huge 40lb wheel of cheese. People would come in for a cup of tea and hack great lumps of the concoction, while others would wander in and buy cheese direct from the farmhouse kitchen.

To Lucy Appleby cheese-making was an art. She argued that cheese was a living thing in which the natural bacteria enhanced the flavour and texture of the final product; to pasteurise it was to kill its character.

As for wrapping it in wax, she helped to persuade makers in neighbouring Lancashire to revert to using calico.

Today Appleby's sells more than one and a half tons of the cheese a week. Customers include Harrods, Selfridges and Fortnum & Mason, and the cheese is exported to the United States, Singapore and the Caribbean.

In 2000 Lance and Lucy Appleby were jointly appointed MBE.

Lance Appleby died in 2003, and their fifth child, Robert, in 2002. Four daughters and two sons, one of whom (Edward) now runs the business, survive her.[1]

Why would Lucy Appleby devote her professional life to defending raw milk cheesemaking in England? Why did she refuse to adopt industrial cheesemaking technologies? And why would prestigious purveyors, such as Harrods, Selfridges, and Fortnum & Mason continue to sell Appleby's raw milk cheese internationally if it were unsafe?

There are two vastly distinct worlds of cheese. Most Americans' experience with cheese is limited to cheese as an ingredient: mozzarella topping a hot pizza; shredded cheese in tacos, nachos, or other Mexican-style foods; cream cheese on a bagel; Cheddar cheese in comfort foods such as macaroni and cheese. Given consumer demand for fast food and prepared meals, the processed food industry is a major user of cheese as an ingredient. The requirements in the processed food industry dictate that cheese function as a perfectly predictable ingredient: that it melt with consistency, brown in a predictable manner, and appear as a perfect emulsion so that oil does not separate from the cheese mass. As a result, cheeses used as ingredients often contain products (processing aids) that help them behave consistently. There are many wonderful processed cheese products that have created great demand for milk and milk-derived ingredients, such as milk protein concentrates. This is the world of the processed cheese industry, where cheese is made with ingredients and technologies that have attempted to improve upon traditional practices of the past to speed up ripening, improve flavor, and create products in large volume quickly to meet consumer demand worldwide at a price point designed to appeal to the masses.

There is another world of cheese, though; one that shares rich cultural roots and traditions with varieties that have been manufactured in Europe for centuries. These age-old cheese traditions are now being embraced by US artisan cheese producers and, as a result, a bona fide artisan cheese industry has emerged in the United States. These cheeses vary in consistency, depending on what starter culture was used, the type of bacterial flora present in the milk, the aging conditions, the pastures or forages on which the animals were raised, and the animals used to produce the milk, ranging from sheep to goats to cows to water buffalo. These cheeses are living, breathing entities, and the appeal of these products is their diversity, their uniqueness, and their lack of uniformity and predictability. They have in some cases a very short shelf life. Like fine wines, artisan cheeses have a terroir, where subtleties of flavor reflect the soil, pastures, and microclimate of regional production. It is the search for uniqueness and inconsistency that connoisseurs of artisan cheeses crave. These two worlds of cheese seem to coexist quite nicely in Europe, and together create an exciting range of foods that consumers enjoy. In the United States, these two worlds collide, particularly when the topic of raw milk cheese becomes inserted into the equation.

This book illustrates the raw milk cheese debate in all its complexity. Cheese is a food that has been produced and consumed for centuries, with recent archaeological evidence dating cheesemaking as far back as 8000 BCE.[2] Many of the world's great traditional cheeses originated in Europe, where they have been continuously produced for 700 years or longer. Despite being made from raw milk, these products enjoy an enviable track record of food safety, largely due to investments in scientific research made by the European Union to protect its traditional foods. Traditional cheeses are products that are highly prized by consumers, and consumer interest in artisan cheese has never been higher.

It is therefore curious that, particularly here in the United States, over the past 20 years, the FDA has waged an unrelenting war against the artisan cheese community, putting in place regulations aimed at eliminating the production of cheese made from raw milk. These efforts have negatively impacted artisan cheese production in the United States and restricted importation of cheeses produced outside the country. While the FDA has claimed that regulatory activity was necessary to ensure food safety, I will illustrate that this is not at all the case. While the FDA claims it is a science-based agency, the science supporting the FDA's efforts at artisan cheese regulation is at best questionable, and at worst incorrect. As we explore this issue, you will understand why it became necessary for people like Lucy Appleby, HRH Prince Charles, Mateo Kehler, Cathy Strange, Giuseppe Licitra, Sylvie Lortal, and countless other individuals to defend the merits of raw milk cheesemaking and thereby preserve important food traditions here and abroad.

My personal introduction to the raw milk cheese debate came on September 20, 2000, when I received an unsolicited letter from a colleague who was working as a consultant to the Cheese of Choice Coalition (CCC). The CCC had as its member organizations the Cheese Importers Association of America, Oldways Preservation Trust, the American Cheese Society, the Parmigiano Reggiano Consortium, and Whole Foods Market, among others. The coalition was developing a body of scientific evidence to help inform the FDA in its decision-making regarding the use of raw milk in the production of aged cheeses. I had been recommended as an expert who could help perform a literature search on raw milk cheese safety, the epidemiology of raw milk cheese outbreaks, and the production of cheese and its safety worldwide. I was asked whether, under certain conditions, the use of raw milk in the manufacture of cheese is appropriate, and was asked to reply to the inquiry.

That letter sat prominently on the corner of my desk for three months. I think I read the letter daily, thinking about the scientific, political, and professional implications of my joining such an effort. Each time I picked up the letter, I could foresee the potential controversy and political conflict associated with the issue, and I would set the letter aside. Dairy politics are fierce, and as a scientist I have tried my best to remain apolitical on scientific issues. My expertise is in dairy product microbiological safety, and I have spent almost 35 years working to protect public health by promoting dairy product safety. I joined the faculty of the University of Vermont in 1983, at the very time an outbreak of listeriosis had infected 49 people and resulted in 14 deaths in Boston. The infective vehicle was later determined to be pasteurized milk, with the raw milk having been produced in Vermont and transported to Massachusetts for pasteurization and processing. I began research on *Listeria* at that time, and it has been the focus of my career ever since.

During the investigation of the Boston outbreak, the US Centers for Disease Control and Prevention (CDC) had developed an operating theory that *Listeria* may have survived the pasteurization process, leading to the contamination of the milk and consequent illness of some persons consuming this contaminated milk.[3] Pasteurization has been the bedrock of the dairy industry, an industry unequivocally committed to public health. I spent three years working collaboratively with the FDA on research that ultimately proved that pasteurization was effective in destroying *Listeria*, and that the most likely route for milk contamination was post-pasteurization recontamination from the processing environment. We now recognize *Listeria* as an important environmental contaminant of food processing facilities, and each year the US food industry spends vast sums of money to control *Listeria* in food processing plant environments.

As I read and reread the Cheese of Choice Coalition letter, I began my own investigation on the safety of raw milk cheese. The FDA, in an April 1997 referral to the National Advisory Committee on Microbiological Criteria for Foods (a committee that I would join in 1999), asked if a revision of policy requiring a minimum 60-day aging period for raw milk hard cheeses was necessary. The FDA, in its communication, noted that such a duration might not be sufficient to provide an adequate level of public health protection. The policy revision being contemplated was a mandatory requirement for pasteurization of all milk used for cheesemaking. The mandatory use of pasteurized milk in cheesemaking was also being considered by Canada and the UK in 1996. In the United States, the legal options cheesemakers have are to use pasteurized milk or, for certain specified cheeses, to hold cheese made with raw milk for 60 days. By default, eliminating the 60-day holding option would leave only the option of using pasteurized milk in cheesemaking.

In response to this proposed FDA action, groups had formed on polar extremes of the issue. On December 12, 2000, I finally had developed both the scientific interest and resolve to pick up the phone and agree to work with the Cheese of Choice Coalition and thereby publicly join the raw milk cheese debate. The FDA and other scientific bodies have stated that the raw milk cheese debate is about infection and contamination control. I will argue that this is not at all the case. The raw milk cheese debate is really a debate over where and how our food is produced and by whom, the values that we individually place on methods of food production, and the conflicting roles of tradition, heritage, and quality versus advertising, marketing, politics, and profits, which ultimately influence our food choices.

American Artisan Cheese Comes of Age

In considering the great cheeses of the world, and questioning which country has the world's greatest cheese culture, it would be hard to argue that the answer is not France, where traditional cheeses have been produced and enjoyed for centuries. Imagine the quandary when, on February 11, 2014, President Obama hosted French President François Hollande as the guest of honor at a White House state dinner. What cheese or cheeses would the White House executive chef select to serve to the leader of the greatest cheese culture on earth? Which of the 400 named French cheese varieties would be showcased at the state dinner? The answer was, none of these.

Instead, featured prominently on the menu as part of the main course was Jasper Hill Farm's blue cheese, produced in my own home state of Vermont in the tiny town of Greensboro, a place near and dear to my heart.[4] That moment cemented for me the rise to prominence of American artisan cheese and the establishment of a bona fide cheese culture in the United States. This cheese, Bayley Hazen Blue, produced by the Cellars at Jasper Hill Farm, was named the World's Best Unpasteurized Cheese later that year on November 16, 2014, at the World Cheese Awards in London. The international acclaim enjoyed by the Cellars at Jasper Hill Farm and many other US artisan cheesemakers reflects trends in the US artisan cheese industry, where states such as New York, Maine, Pennsylvania, California, Wisconsin, Washington, and Vermont lead the country in artisan cheese production. Other states such as Michigan, North Carolina, and Texas, which have not traditionally had artisan cheesemaking, are seeing tremendous increases in artisan cheese production. Demand for these locally produced goods outpaces supply, and these products are creating new and vibrant economic opportunities in rural communities.

Why, then, has the US artisan cheese industry been subjected to so much regulatory scrutiny? Over the past 20 years, the FDA has directly challenged both domestic artisan and imported cheeses. Issues such as the 60-day aging rule, the soft cheese risk assessment, the *Listeria* swabbing assignment, the use of wooden boards for cheese aging, the use of ash on cheese surfaces, and the establishment of stringent *E. coli* standards are but a few of the challenges that have arisen aimed at the artisan cheese industry, both domestically and globally. The common denominator between many US artisan and imported cheeses is that they are produced from raw milk.

To fully understand these issues and place them in context, it is necessary to explore the history of raw milk cheese production in the United States and why raw milk cheesemaking is legal. My dear friend and colleague Professor Paul Kindstedt, in his acclaimed book *American Farmstead Cheese*, provides an excellent historical account of the Old World origins of cheese and how these traditions were brought to the New World.[5] In particular there were the Puritans, who migrated from East Anglia, a well-established cheesemaking region in England. With them they brought their cheesemaking knowledge and expertise that allowed the production of durable English hard cheeses like Cheddar. By 1650, cheesemaking was widespread in New England, and this trend would continue through to 1840, when Cheddar production dominated US cheesemaking. A USDA map from 1848 showing cheese made on farms confirms that cheesemaking was common throughout New England and the Middle Atlantic states and would flourish until two key events reshaped cheesemaking in America. The first was the development of railroads, which allowed fluid milk intended for pasteurization to be shipped great distances. The second occurred in 1851, when Jesse Williams and his son George applied the concept of industrialization to cheesemaking, a move that radically altered the US cheesemaking landscape. From this point onward cheese factories were constructed in New York State and elsewhere, shifting cheese production off the farm. By the 1940s large-scale industrialized cheesemaking had replaced on-farm artisan cheesemaking, with Cheddar cheese being the predominant cheese produced in the United States.

In 1941 with the advent of World War II, the US Congress passed the Lend-Lease Act, which resulted in a government purchase of 150 million pounds of US Grade 1 Cheddar Cheese. The purpose of this purchase was to supplement the diets of US armed forces personnel. This large government purchase required increased production of cheese. To respond to the production demands, unskilled workers began making cheese using obsolete equipment, with improper temperature controls and holding times, and thus did not control the quality of the products that were produced. These practices led to typhoid, brucellosis, and staphylococcal outbreaks linked to Cheddar cheese during the mid-20th century in the years surrounding World War II.[6] In response to these outbreaks, the US Food and Drug Administration published two options for producing safe cheese products.[7] Cheesemakers could either pasteurize milk intended for cheesemaking or they could hold cheese at a temperature of not less than 35°F (2°C) for a minimum of 60 days. The latter option became known as the 60-day aging rule. This regulation remains part of Title 21 of the US Code of Federal Regulations (CFR) Part 133 to this day.

Why does the 60-day aging rule promote cheese safety? The regulations were formulated around the behavior of bacterial pathogens such as *Salmonella typhi* and *Brucella* in Cheddar cheese, the predominant cheese variety produced in the United States in 1950. Cheddar cheese has a low moisture content, high salt content, and low pH / high acidity, and these parameters interact to create an environment that is inhospitable to bacterial pathogens, so they die off as cheese ages over the course of 60 days or longer. Not all cheeses share these characteristics, however, and the regulations currently upheld in the CFR have been broadly applied to a number of specified cheese varieties despite scientific evidence that suggests such regulations are inappropriate for certain cheeses, such as soft-ripened varieties like Brie and Camembert.

Do cheese regulations promulgated in 1950 serve to protect public health today? To a large extent, the answer is yes, as confirmed by comprehensive reviews published in the scientific literature. If cheesemakers elect to use pasteurization of milk for cheesemaking, this heat treatment process kills bacterial pathogens in milk. It helps to ensure consistency and quality of cheeses. If the raw milk used for cheesemaking is of inferior quality, pasteurization of this milk may improve flavor, cheese yield, uniformity, and shelf life, and simplify the cheese manufacturing process. Importantly, pasteurization of milk was found to extend the shelf life of cheese when it was being shipped across great distances. Brie cheese was first imported to the United States from France in 1936. Pasteurization was used to facilitate the export of French Brie because of the need to find a stable and safe way to distribute this cheese.

Alternatively, cheesemakers can hold some cheese varieties for 60 days or longer, and in certain cheese varieties pathogens will die off during 60 days of aging. Yet we began to see challenges to the 60-day aging rule, particularly in the 1990s. Why at that particular time? We will get to that point later in this book. For now, it's enough to note that a number of studies were conducted and published in the mid-1990s showing survival of dangerous pathogens such as *Salmonella typhimurium*, *E. coli* O157:H7, and *Listeria monocytogenes* in cheese beyond the mandatory 60-day holding period. The FDA referred this issue to the National Advisory Committee on Microbiological Criteria for Foods (NACMCF) in 1997. The FDA asked this scientific advisory committee if a policy revision was necessary, indicating that the 60-day rule might not be sufficient to protect public health. Around the same time period, the Institute of Food Science and Technology (IFST) in the UK had issued a position statement citing potential health hazards due to the presence of pathogens in raw milk cheese.

What would happen if the 60-day aging rule were to be changed and cheesemakers were forced to pasteurize their milk used for cheesemaking? Of most importance would be the impact on global trade and the importation of aged hard Swiss, Italian, and Cheddar cheeses. In fact, as we will see, it is likely that global trade, and not food safety, is really at the heart of the raw milk cheese debate.

CHAPTER 3

Is Raw Milk Cheese Really Risky?

As global demand for cheese continues to grow, so will the need to continue to ensure that cheese is safely produced. Cheese safety is accomplished either by adherence to time-honored artisan practices that process raw milk, including heat treatment of curd, aging of cheese to facilitate die-off of harmful organisms, or, alternatively, through the use of pasteurized milk in cheesemaking. Is there a need to change the 60-day aging rule due to food safety concerns? When scientists are asked to explore this question, we usually turn at first to the scientific literature to review what has been published on this topic. So what does the scientific literature have to say about cheese safety? A comprehensive, three-part 1990 publication by Dr. Eric Johnson and his colleagues at the University of Wisconsin reviewed the epidemiological literature during the 40-year time period from 1948 to 1988.[1] Only six outbreaks of illness occurred during this entire time frame from cheese that was manufactured in the United States, despite the fact that 100 billion pounds of cheese were produced and consumed during these years. Interestingly, post-pasteurization recontamination was cited as the most frequent cause of the outbreaks. The authors also reviewed cheese-related outbreaks occurring in Canada between 1970 and 1987. Only two outbreaks that occurred, one each in the United States and Canada, involved cheese made from raw milk. No outbreaks during this 40-year period were linked to Italian hard cheese varieties (Parmesan, Romano, and Provolone). Factors other than use of pasteurization that were found to contribute to cheese safety included the following: milk quality and management; lactic acid bacteria starter culture management and bacteriophage control; pH, salt, use of controlled aging conditions; and the presence of natural inhibitory substances in milk.

A 1998 publication by authors working for the Centers for Disease Control and Prevention (CDC) reviewed all cheese-associated outbreaks reported to the CDC between the years 1973 and 1992.[2] The authors reported that improved cheesemaking methods and process controls resulted in cheese being a safer product. The authors noted 32 cheese-associated outbreaks that occurred during this period. Of these, 11 were attributed to contamination at the farm, during manufacturing, or during processing. Of the 11 outbreaks attributed to contamination prior to distribution, 5 were associated with the consumption of "Mexican-style soft cheese" (*queso fresco*). Curiously, no outbreaks reported to the CDC during the years 1973 to 1992 were associated with raw milk cheeses that had been aged for 60 days. The CDC concluded in this article, "We reviewed all cheese-associated outbreaks of human illness reported to the CDC with onsets during 1973–1992. The infrequency of large, cheese-associated outbreaks was notable."

A 2008 review reported cheese-related outbreaks that had occurred in the United States, Canada, and Europe during the years 1980 to 1996.[3] Of 30 outbreaks reviewed by the authors during this period, 16 outbreaks were linked to cheese made from raw milk, while 14 were linked to cheese made from pasteurized milk. Pathogens involved in the outbreaks included *Brucella*, *E. coli*, *Listeria monocytogenes*, *Salmonella*, and *Yersinia enterocolitica*. Three outbreaks due to *L. monocytogenes* were responsible for 284 illnesses and 88 deaths. The authors noted that since 1996, periodic outbreaks of *L. monocytogenes* have continued to occur.

So based on these reviews, it would seem that outbreaks of illness linked to cheese are infrequent, but when they do occur, they can be traced either to cheese made from unpasteurized milk or to cheese made from pasteurized milk that is recontaminated after pasteurization. Further, in outbreaks involving cheese made from unpasteurized milk, it often comes from cheese that is illegally produced in unlicensed facilities, and a certain type of cheese, *queso fresco*, predominates in the outbreaks. Why is *queso fresco* a problem? This traditional Mexican cheese is often illegally produced, using raw milk. It is made without the use of a starter culture, it is high in moisture, and the pH remains within ranges that can support the growth of pathogens such as *Listeria* and *Salmonella* to high levels. This is not a cheese that is aged for 60 days; it is consumed fresh. To ensure its safety, it must be made from pasteurized milk. It is therefore curious that the FDA and others push for changes to the 60-day rule when the outbreaks involving *queso fresco* are cited. Eliminating the 60-day aging rule will not stop the illegal production of raw milk *queso fresco*. Further, mandating pasteurization of all milk for cheesemaking will not prevent outbreaks linked to recontamination of cheese following pasteurization.

Challenge Studies

In addition to literature reviews, when concerned about the behavior of pathogens in a food product, experiments called *challenge studies* can be conducted to determine if growth or survival of a pathogen can occur. The FDA justified its reexamination of the 60-day aging rule based on limited scientific challenge studies, some of which it conducted internally. In 1992, there had been a major outbreak of illness in the United States caused by the pathogen *E. coli* O157:H7 in undercooked hamburgers served by the Jack in the Box restaurant chain. A total of 732 individuals were sickened and 4 children died.[4] Unlike normally harmless strains of *E. coli*, this particular strain, not seen before 1982, had acquired a gene that allowed for the production of cytotoxins, or Shiga toxins, which can be deadly, especially for children. The reservoir in nature of Shiga toxin–producing *E. coli* (STEC) is dairy cattle. Only a few cells are needed to cause illness, and infection can lead to permanent damage of kidney function in young children. Following this tragic outbreak, challenge studies were conducted to examine the behavior of *E. coli* O157:H7 in a number of foods, including cheese. A study published in 1996 examined the survival of *E. coli* O157:H7 during the manufacture and ripening of Cheddar cheese.[5] Milk for cheesemaking was inoculated with *E. coli* at two different levels of colony forming units (cfu)—1,000 cfu/ml (treatment 1) and 1 cfu/ml (treatment 2). In both treatments, a sharp decline in *E. coli* was noted, but in treatment 1, *E. coli* was detected after 158 days of aging. In treatment 2, *E. coli* was undetected at 60 days of cheese aging. The study suffered from a number of problems in its experimental design. First, the salt level in the study cheese (as measured by salt in the moisture phase; SMP) was 2.75–3.76; normal Cheddar cheese has an SMP of 5–5.5 percent. As salt is an important inhibitor of pathogen growth, the low salt levels in the study cheese may have contributed to *E. coli* O157:H7 survival. Second, the study used pasteurized milk instead of raw milk. Why would you use pasteurized milk if you are trying to study what happens to *E. coli* in raw milk cheese? In fact, this is a common problem in many scientific studies that seek to examine the fate of pathogens during raw milk cheese manufacture. Third, the *E. coli* levels used in treatment 1 are substantially higher than what would normally be found in raw milk. Despite the study flaws, the FDA often cites this study as scientific evidence suggesting that the use of raw milk in cheesemaking should be banned and the 60-day aging rule revised.

The FDA conducted its own internal studies that were published in 2006.[6] This time, the FDA's experiments did include the use of unpasteurized milk. The raw milk was inoculated with three levels of *E. coli* O157:H7: 100,000 (10^5); 1,000 (10^3); and 10 (10^1) cfu/ml. The investigators explained that the highest inoculation level (10^5) was needed for enumeration purposes during the experimentation, although the authors did indicate that 10 cfu/ml was the level of *E. coli* O157:H7 expected to be encountered in raw milk "in the real world." Again, there were major flaws with this study design. The high *E. coli* levels used in this study would raise concerns about the quality of the raw milk used for cheesemaking, as they exceeded FDA's legal requirements for raw milk. The cheese also contained *E. coli* levels that exceeded the FDA's legal requirement for cheese. In Guide 7106.08, published August 1, 1986, the FDA had established standards for enteropathogenic *E. coli* EEC (10^3; 1,000) and *E. coli* (10^4; 10,000) in cheese. The cheese did not conform to these standards, although this fact was never noted in the publication. In fact, in its published study, initial populations of *E. coli* in cheese reached 1.6×10^7 / ml. Despite the study flaws, populations of these organisms were found to decline over time. However, the authors concluded in their publication: "These studies confirm previous reports that show 60-day aging is inadequate to eliminate *E. coli* O157:H7 during cheese ripening."

The category of cheeses posing the greatest potential risk to consumer health and safety are the soft and semi-soft cheeses, including soft bloomy-rind cheeses such as Brie and Camembert, soft natural-rind cheeses such as Tomme style, and soft cheeses washed in brine (red smear cheeses including

Taleggio and Muenster). A number of outbreaks of illness have been linked with the consumption of soft and semi-soft cheeses. A large outbreak of listeriosis occurred between the years 1983 and 1987 that was linked to consumption of Vacherin Mont d'Or cheese, a cheese made in Switzerland from thermized milk, even though the outbreak is frequently referenced as one involving a cheese made from raw milk.[7] Instead of banning the production of Vacherin Mont d'Or, an appeal was made by the physician investigating the outbreak, Dr. Jacques Bille of the University Hospital of Lausanne and former head of the Swiss National Reference Center for *Listeria*. In an impassioned presentation of outbreak data at a meeting that I attended, Dr. Bille stated, "We must save this magnificent cheese." Implementation of rigorous cleaning and sanitation protocols in caves and manufacturing environments along with use of equipment that could be effectively cleaned and sanitized was extremely effective in ending the outbreak. As will be discussed, the Swiss government has invested extensive scientific resources to prevent *Listeria* contamination in cheese-aging facilities, as this is the best way to ensure cheese safety.

Illegally produced *queso fresco*, the soft fresh Hispanic-style cheese (frequently referred to as Mexican-style cheese in government publications) discussed previously, was linked to outbreaks of listeriosis in California in 1985 (described below) and North Carolina between 2000 and 2001, as well as an outbreak of *Salmonella* Newport, in California in 2002. When produced legally in inspected commercial facilities, *queso fresco* is a very safe product. However, it is often produced by unlicensed manufacturers under insanitary conditions, leading it to be referred to by regulators as *bathtub cheese*. Back in 2005, the FDA issued an advisory about certain soft cheeses made from raw milk. Raw milk soft cheeses were associated with cases of listeriosis, brucellosis, salmonellosis, and tuberculosis. *Queso fresco* from Mexico, Nicaragua, and Honduras was of most concern. These cheeses were transported in luggage, sold door-to-door by unlicensed vendors or sold in supermarkets. *Queso fresco* from New York City was linked to a case of tuberculosis. Unfortunately, the FDA's concerns about this fresh, soft, and often illegally produced raw milk cheese raised concerns about *all* cheeses made from raw milk.

The outbreak in Los Angeles, California, in 1985 was caused by Jalisco brand cheese. A total of 142 cases of illness and 48 deaths were linked to this outbreak. The majority of cases involved pregnant women and their fetuses. FDA inspection records revealed that the company that produced the contaminated cheese used a combination of raw and pasteurized milk during manufacturing, and an inspection of records revealed that the volume of cheese produced by the plant greatly exceeded the capacity of the pasteurizer. Despite the size of this outbreak, it is notable that it would have gone undetected had it not been for a study on causes of spontaneous abortion and stillbirths in pregnant Hispanic women that was being conducted at hospitals within the Los Angeles Department of Health Services network. At the time, this was the largest birthing center in the United States for pregnant Hispanic women.[8]

For soft and semi-soft cheeses, post-pasteurization recontamination (contamination that occurs after pasteurization) has been identified as a factor in several outbreaks. A 2001 study examined the incidence of *L. monocytogenes* in European red smear cheeses.[9] The authors found a higher incidence of *L. monocytogenes* in cheeses made from pasteurized (8 percent) versus raw (4.8 percent) milk.

Curiously, despite the FDA's attempts to ban all raw milk used in cheesemaking and change the 60-day rule, it is actually the FDA's own regulations that create food safety problems for cheesemakers. The US Code of Federal Regulations (CFR) Title 21 addresses regulations for food to which companies must adhere. Part 133 states requirements for cheese. CFR 133.182 is a section dealing with soft-ripened cheeses, and CFR 133.187 addresses semi-soft cheese. In both cases, the 60-day aging rule applies to these cheeses. These regulations show the contrast between our regulations and those in Europe. A soft-ripened cheese such as Camembert was never meant to be aged for 60 days, and holding this cheese for that time period only increases, not decreases, the microbiological risk, especially for *Listeria*. In Cheddar cheese, *Listeria* dies during 60 days of aging due to the combination of acidity, salt, and low moisture that develops during aging. In contrast, the pH of Camembert cheese increases during ripening as a result of the mold growth, or bloom, on the cheese surface. The high moisture content and neutral pH of Camembert cheese promote the growth and survival of *Listeria* during ripening, with populations reaching dangerous levels at 60 days of ripening. This cheese is at its peak of perfection at 30 days of ripening, and in France it cannot be sold after 55 days, because the microbiological risk is too high. Why the 60-day aging rule applies to soft-ripened cheese in CFR 133.182 is unknown.

In 1995, Canada had a proposal to require all cheeses to be made from pasteurized milk or the equivalent. The European Communities (EC) threatened a World Trade Organization (WTO) dispute if Canada proceeded with the raw milk cheese ban.[10] In an appeal to the WTO, the EC raised concerns with the proposed Canadian requirement. The EC considered that its cheesemaking measures, including production requirements, safe and correct sourcing of milk, and subsequent supervision in the various production stages from farm to consumer, provided at least equivalent guarantees of food safety when compared to cheeses made from pasteurized milk. The European Communities observed that a number of WTO members maintained restrictions on imports of raw milk cheeses from the EC, which were not justifiable on health grounds. Canada noted that, on request, it would provide scientific documentation in support of its proposal, and that a scientific expert advisory committee had been appointed to examine the matter.

Canada ultimately withdrew this proposal in 1996 when its scientific expert committee, sympathetic to the plight of Quebec cheesemakers, concluded that the technical requirements presented by a pasteurization requirement could not be met by all cheesemakers in the cheese manufacturing process. Quebec enjoys a rich cheese culture that includes production and consumption of traditional raw milk cheeses, and cheesemakers cited cultural and economic reasons for opposing the ban on raw milk cheesemaking. They instead placed emphasis on hygienic milk production practices to promote food safety.

As reported by Hornsby:

> By the end of 1996, the Canadian Scientific Expert Advisory Committee on Raw-Milk Soft Cheese reported back on the health risks and whether the pasteurization requirement was necessary. To the surprise of federal officials, the Committee did not advocate for the pasteurization requirement, and instead advocated for a five-step process to ensure that raw-milk cheese for human consumption did not get contaminated with pathogens and when it did, that the source could be traced. Indeed, the

> expert committee even suggested that Health Canada's existing sixty day aging requirement was unnecessary as long as the five step process was in place. With the release of the expert committee's report, the Canadian proposal as a domestic measure was considered no longer defendable against Quebeçois objections.[11]

Even our own US National Academy of Sciences' panel of experts, in the 2003 publication *Scientific Criteria to Ensure Safe Food*, recommended against mandatory pasteurization of milk used for cheesemaking and instead called for "development and implementation of a scientifically appropriate performance standard for the reduction of targeted pathogens in finished cheese products as a consequence of the processing strategies and aging periods employed in the manufacturing of the products."[12]

Despite the vast body of scientific evidence documenting the safety of aged raw milk cheeses, the FDA continues to push forward with its intent to ban raw milk cheesemaking. As recently as August 3, 2015, the FDA published a docket in the Federal Register, seeking public comment on the following notice: "Understanding Potential Intervention Measures to Reduce the Risk of Foodborne Illness from Consumption of Aged Cheeses Manufactured from Raw Milk." Here are portions of the docket request:

> The Food and Drug Administration (FDA or we) is requesting comments and scientific data and information that would assist us in identifying and evaluating intervention measures that might have an effect on the presence of bacterial pathogens in cheeses manufactured from unpasteurized milk. We are taking this action in light of scientific data on potential health risks associated with consumption of cheese made from unpasteurized milk. . . .
>
> More recently, the results of the FDA/Health Canada QRA suggest that the 60-day aging period for soft-ripened cheese may increase the risk of listeriosis from consumption of soft-ripened cheese by allowing more time for L. monocytogenes, if present, to multiply (rather than decrease) as the softripened cheese ages (Ref. 6). . . .
>
> FDA recognizes that there is broad diversity in cheese manufacturing operations and approaches and that many factors go

> into ensuring the safety of the food. Many types of raw milk cheeses are made using traditional methods that require a successful balance involving the quality of the milk, the equipment, and the environment, including ensuring the presence of bacteria critical to the nature of the cheese while preventing the introduction or growth of pathogens. In issuing this call for data and information, we are particularly interested in learning more about the standards and practices in use by the growing artisanal cheese manufacturing community. . . .[13]

The FDA's challenge to the safety of cheese, as stated in the request for scientific data and information, is misleading, as it disregards the long track record of safety that legally produced cheese has enjoyed. Much of the challenge to cheese safety in the FDA's request for scientific data documentation comes from a CDC study conducted by Langer and others in 2012,[14] which was referenced in the FDA's call for data. As stated in the docket:

> A 2012 review of outbreaks of foodborne illness that occurred in the United States between 1993 and 2006 that were attributed to dairy products determined that more than 50 percent of the outbreaks reviewed in the study involved cheese, with the remaining outbreaks being attributable to fluid milk (Ref. 1). Forty-two percent of the 65 cheese-associated outbreaks (i.e. 27 outbreaks) were attributable to products manufactured from unpasteurized milk, even though the contribution of unpasteurized dairy products to all dairy product consumption in the United States during the time period under study was estimated at below 1 percent (on a weight or volume base). The 65 analyzed outbreaks due to cheese resulted in 641 associated illnesses with 131 hospitalizations) (i.e., a hospitalization rate of more than 20 percent). Pathogens associated with these outbreaks included *Listeria monocytogenes*, *Escherichia coli (E. coli)* O157:H7, *Salmonella*, and others.

The Langer study did not consider the type of cheese made from unpasteurized milk, which is a critical consideration for the identification of effective mitigation strategies to control pathogens in cheese. However, 27 outbreaks occurring over a 13-year period reveals that only 2 outbreaks per year are linked to cheese made from unpasteurized milk. This would suggest an excellent track record of safety and suggests that current regulations appear to be working as intended to protect public health. As noted in the Langer information above, if 42 percent of cheese outbreaks involved cheese made from unpasteurized milk, then the remaining outbreaks must have involved cheese made from pasteurized milk. Therefore, mandating pasteurization of all milk for cheesemaking will not prevent outbreaks of illness from occurring. A review of the references provided by FDA in its docket shows that a majority of cases of illness attributed to cheese made from unpasteurized milk as documented by Langer and others were due to one particular cheese type: *queso fresco*. As previously discussed, the dangers of this type of cheese have been well documented by the FDA, CDC, and others. The Langer study references an article by Headrick that explores the epidemiology of raw milk associated foodborne disease outbreaks that occurred between the period 1973 and 1992.[15] The outbreaks discussed by Headrick are linked to fluid raw milk (not raw milk cheese) consumption. Curiously, the Langer study fails to cite the CDC article, authored by Altekruse, discussed previously on page 24. In fact, the title of the Altekruse article is "Cheese-Associated Outbreaks of Human Illness in the United States, 1973 to 1992: Sanitary Manufacturing Practices Protect Consumers."

The Langer article references work by Sara H. Cody, who reported on two outbreaks of illness due to consumption of Mexican-style fresh cheese that occurred in Northern California.[16] In the first outbreak, cheese had been purchased at a flea market. A week later, a second outbreak occurred involving Mexican-style cheese purchased from street vendors. The California Department of Agriculture traced the cheese production to a backyard shack, where a garden hose was the only source of water available for equipment cleaning. The dangers of this type of high-risk cheese, being manufactured and distributed by unlicensed producers and unlicensed vendors, remains a public health problem. This type of cheese production and distribution should not be confused with artisan cheese made in inspected and licensed facilities under highly sanitary conditions.

Langer also includes a reference to an article by Méndez Martínez.[17] This article describes an outbreak of brucellosis that occurred in Spain and was linked to unpasteurized goat cheese. In citing this article, Langer and others fail to note that the cooperative state-federal Brucellosis Eradication Program was launched by the USDA Animal and Plant Health Inspection Service (APHIS) in 1954. Today, dairy calves, cattle, and domestic bison in all 50 states, Puerto Rico, and the US Virgin Islands are brucellosis-free.

The FDA docket importantly fails to reference a follow-up study conducted by L. Hannah Gould of the CDC. The article, titled "Outbreaks Attributed to Cheese: Differences between Outbreaks Caused by Unpasteurized and Pasteurized Dairy Products, United States 1998–2011," documents 90 outbreaks occurring between 1998 and 2011 in which cheese was a vehicle of infection. Thirty-eight outbreaks (42 percent) involved cheese made with unpasteurized milk while 44 (49 percent) involved cheese made with pasteurized milk. As stated by the author:

> The most common cheese–pathogen pairs were unpasteurized *queso fresco* or other Mexican-style cheese and *Salmonella* (10 outbreaks), and pasteurized *queso fresco* or other Mexican-style cheese and *Listeria* (6 outbreaks). The cheese was imported from Mexico in 38% of outbreaks caused by cheese made with unpasteurized milk.[18]

As previously discussed, *queso fresco* is often made in unlicensed facilities and distributed by unlicensed vendors, or illegally imported from Mexico and Central America. Placing burdens on the legal/licensed cheesemaking community, such as the establishment of performance standards or a requirement for additional testing and procedures to eliminate pathogens in unpasteurized milk, will not address the problem of illegal production and illegal importation of cheese.

This problem is further exacerbated by changing state regulations, which are allowing increased sales of unpasteurized milk. Consumption of unpasteurized fluid milk represents a known public health problem, and cases of illness will continue to appear due to raw milk sales and consumption. It is important, however, to recognize the differences in the microbiological risk profile of fluid raw milk versus the microbiological risk profile of raw milk cheese. With fluid raw milk consumption, *Campylobacter* is a frequent cause of illness. In fact, my home state of Vermont has the second-highest burden of *Campylobacter* infection in the United States, and our state epidemiologists attribute this burden to fluid raw milk consumption. In contrast, *Campylobacter* is not a pathogen of concern in raw milk cheese. As documented by Food Standards Australia New Zealand (FSANZ) in its comprehensive "Microbiological Risk Assessment of Raw Milk Cheese" published in 2009, it states:

> *Campylobacter* spp. were found to be a negligible risk in both raw milk extra hard and Swiss-type cheeses. . . . *Campylobacter* spp. are unlikely to grow in milk or cheese as their growth requires reduced oxygen tension and temperatures between 32–45C [90° to 114°F] and they do not survive under slightly acidic conditions or in the presence of greater than 2% salt.[19]

In further reviewing the Gould data, of 44 outbreaks (49 percent) linked to cheese made from pasteurized milk, 64 percent of these outbreaks were linked to cheese served at a restaurant, deli, or banquet with food worker handling implicated with the illnesses. None of the strategies being considered by FDA (control measures, 60 days of aging, other treatments in addition to pasteurization, and performance standards) will protect public health if ill workers are improperly handling products in food service and using cheese as a vehicle to spread contamination from norovirus and other pathogens. Again, the mitigation strategy that is needed involves having food service workers comply with Food Code regulations. This is a compliance issue, and licensed cheesemakers who produce microbiologically safe products should not be subjected to regulatory scrutiny due to lapses in safe food handling by food service workers.

The dangers of cheese cross-contamination at retail when cheeses are cut and wrapped or sliced has been addressed in recent publications. Such contamination complicates epidemiological investigations to identify the source of contamination in illness outbreaks. Application and maintenance of good hygienic practices are needed throughout the food chain in order to prevent contamination and minimize pathogen growth. As an example, imported Frescolina Marte brand Ricotta Salata cheese, a soft cheese made from pasteurized milk, was identified as the cause of a complicated outbreak of listeriosis that occurred in the United States in 2012. The outbreak affected 22 individuals in 13 states. Investigations began in Pennsylvania, where a patient who contracted listeriosis had consumed two soft cheeses purchased from a grocery store: a commercially produced blue cheese made from unpasteurized milk and an imported l'Edel de Cléron cheese made from pasteurized milk. Investigators developed a theory early in the investigation that an intact contaminated cheese could cross-contaminate multiple cheese types during cutting and wrapping. The outbreak strain was isolated from samples of cut and repackaged cheese from both a cheese distributor and a grocery chain. The distributor did not ship the cut and repackaged cheese to the grocery chain, and the grocery chain received only intact wheels. Epidemiological investigations revealed that blue and farmstead cheeses that were cut and repackaged by the distributor were contaminated with the epidemic strain of *L. monocytogenes*, but intact wheels of blue and farmstead cheese did not contain *L. monocytogenes*. Cutting records at the distributor revealed that Ricotta Salata was the only common cheese used at cutting stations for the blue and farmstead cheese. At the grocery store, it was likely that Frescolina Marte brand Ricotta Salata cross-contaminated the blue cheese and l'Edel de Cléron bought by the Pennsylvania patient. The outbreak illustrates the risks of cross-contamination posed by contaminated cheese, and it illustrates the need for the use of validated disinfection protocols and sanitation of wire cutters, cutting boards, knives, and utensils following cutting and wrapping of cheese blocks.[20]

Consumers who wish to purchase *queso fresco* should be encouraged to do so from licensed and inspected manufacturers. I have personally visited licensed Mexican-style cheese production facilities around the United States, and they are some of the best food processing plants I have ever visited; they produce safe products. Encouraging consumers to select safer cheese alternatives within the Mexican-style cheese category is an educational task. Establishing performance standards, requiring testing, and so on will not address this issue.

The FDA's own compliance data document the problems associated with Mexican-style cheeses and identify the main pathogens of concern in domestic and imported cheeses. In an analysis of the FDA's Domestic and Imported Cheese Compliance Program results for the years 2004–2006, 42 out of 2,181 (1.9 percent) imported cheese samples tested positive for *L. monocytogenes*, compared with domestic cheese samples where only 10 of 2,181 (0.45 percent) samples were positive.[21] Approximately half of the positive imported cheese samples (52 percent) were traced to Mexican-style soft cheese, or soft-ripened cheeses from France and Italy. Instead of a performance standard to replace the 60-day aging rule, producers of raw milk cheese should be encouraged to routinely test their production environments for presence of *Listeria* in order to ensure that sanitation controls within cheese facilities are working. It is lack of compliance with regulations that threatens the safety of dairy products including cheeses, as the FDA's own data confirm. US regulatory focus should be given to Mexican-style soft cheeses, particularly those produced in Mexico and Central America. Again, changing the 60-day aging rule will not improve the safety of illegally produced Mexican-style cheeses such as *queso fresco*.

Concerns about *Listeria* in soft cheese led the FDA to conduct, in collaboration with Health Canada, a Joint Soft Cheese Risk Assessment in 2012.[22] The risk assessment was conducted due to outbreaks of listeriosis linked to cheese in both the United States and Canada. The Quantitative Microbial Risk Assessment (QMRA) was conducted to evaluate the public health impact of intervention strategies to reduce or prevent *Listeria monocytogenes* in soft-ripened cheese. Results of the risk assessment found that if soft-ripened cheese was made from raw milk and every lot was tested for *Listeria*, this would result in a predicted higher level of safety than use of pasteurized milk in cheesemaking.

In summary, the scientific literature supports the long record of safety that legally produced cheeses have enjoyed. Calls by regulators for mandatory pasteurization of all milk used for cheesemaking on the grounds of public health and safety are not scientifically supported. Why, then, does the FDA continue to focus on the safety of cheeses made from unpasteurized milk?

CHAPTER 4

Why Raw Milk Cheese Is Not Raw

The term *raw milk cheese* is really a misnomer for many cheese varieties. The numerous steps in cheesemaking transform a raw material—milk—and process it for safety through the "hurdle concept" whereby factors such as competition with starter cultures and other bacteria present in raw milk, acidity development, moisture loss, and use of salt all interact to create an environment that is hostile to pathogens. During the process of cheesemaking, the protein in milk, casein, is coagulated to form curds that, in many cheese varieties, receive heat exposure. For some cheeses, even though cheesemaking begins with the use of raw milk, these are cooked-curd cheeses that are anything but raw. Cheeses made from milk subjected to a sub-pasteurization heat treatment (thermization) are legally classified as raw milk cheeses. Thermization is useful for milk that has been transported and stored under refrigeration, or when there will be a delay in the use of milk before cheesemaking. The use of thermization helps to reduce the growth of psychrotrophic bacteria (those that can grow at refrigeration temperatures) that cause quality defects in cheese. Much of the milk used in the United States for raw milk Cheddar cheesemaking is subjected to some form of heat treatment, generally thermization. As a rule, this treatment consists of heating milk at 131°F/55°C for a period ranging from 2 to 16 seconds.

The term *raw milk cheese* also fails to reflect the high curd-cooking temperatures used in the manufacture of aged Swiss and Italian hard cheese

Figure 4.1. Use of copper vats to cook curds during Comté production. *Image courtesy CIGC*

varieties. In fact, many countries refer to these cheeses as cooked-curd cheeses. A 1995 study conducted in Switzerland examined the safety of Swiss hard and semi-hard cheeses.[1] The authors noted that approximately 80 percent of cheeses made in Switzerland are manufactured from raw milk without prior heat treatment. These authors inoculated eight pathogens including *Salmonella*, *Listeria*, and *E. coli* to raw milk at high levels (10,000–1,000,000 cfu/ml). The authors could not detect any pathogens beyond 1 day of manufacture. The microbiological hurdles noted by the authors that contribute to the safety of Swiss hard cheeses include the use of high curd-cooking temperatures (126° to 130°F / 52° to 54°C for 45 minutes); high pressing temperatures (122°F/50°C); low moisture (35 percent); use of high ripening temperatures (50° to 68°F / 10° to 20°C); and use of long ripening times (over 120 days).

Similar hurdles occur during the manufacture of Italian Grana cheeses. A study published in 2001 examined the safety of Italian Grana (Grana Padano and Parmigiano Reggiano) cheeses.[2] The authors noted several parameters that contribute to microbial safety, including cooking and molding of cheese cured at high temperatures (127° to 133°F / 53° to 56°C); brine salting of cheese, which lowers the moisture and water activity of the cheese; and extended ripening of cheese for 9 to 12 months. A rapid decline of pathogens initially added to milk at high levels (100,000 cfu/ml) was observed during manufacture of Parmigiano Reggiano cheese.[3]

In June 1998, Switzerland reported to the World Trade Organization (WTO) that New Zealand and Australia had stopped importing Swiss hard cheeses made from unpasteurized milk because they did not meet established sanitary requirements, which required inactivation of bacterial pathogens.[4] These standards had been established by New Zealand and Australia in January 1995, and compliance with these standards was now being enforced. The WTO reported that Food Standards Australia New Zealand (FSANZ) was evaluating the applications received from Switzerland and the European Communities (EC). In contrast to the FDA's approach to the raw milk cheese issue (that to date continues to be based on limited scientific studies), the Australian and New Zealand governments undertook a series of comprehensive risk assessments to gauge the microbiological safety of hard Swiss cheese varieties (Emmental, Gruyère, Sbrinz) and extra-hard Italian grating cheeses (Parmigiano Reggiano, Grana Padano, Romano, Asiago, Montasio). The results of these risk assessments were published in 2002 and 2003 and established the principle of equivalence.[5] As stated by FSANZ:

> The principle of equivalence in food safety is based on the recognition that the same level of food safety can be achieved by applying alternative hazard control measures. The objective is to determine if these measures, when applied to a food, achieve the same level of food safety as that achieved by applying specified measures.[6]

FSANZ concluded that very hard cheese manufacturing processes can achieve a 5-log reduction of pathogens of concern when using raw milk, given good hygienic and manufacturing practices. Use of a 5-log pathogen reduction process, which is a 100,000-fold reduction in the numbers of the pathogen that is most likely to pose a food safety problem, is a widely used food safety standard. Therefore, the Australian government exempted very hard cheeses of milk pasteurization requirements provided they have a curd heat treatment of at least 118°F (48°C). The concept of equivalence states that these cheeses have an equivalent level of safety to cheeses made from pasteurized milk.

Again, in November 1998, the EC requested through the WTO that Australia identify the international standard on which its import ban on Roquefort cheese was based, or, alternatively, provide scientific justification and a risk assessment.[7] As in its previous response to the WTO, Australia indicated that its food standards required all cheese to be made from pasteurized milk, or milk that had undergone an equivalent process. Australia's preliminary assessment on Roquefort cheese had identified potential problems with pathogenic microorganisms, particularly enterohemorrhagic *E. coli*. Further, data from the Roquefort manufacturers had been received by FSANZ and were being evaluated. In addition to food safety assessments, Roquefort cheese was being evaluated for risks to animal health. The WTO was told that a final decision was likely in the first quarter of 1999 on both food safety and animal health aspects.

FSANZ's scientific evaluation of the safety of Roquefort cheese examined three primary areas: the effect of the cheese manufacturing processes on target microbial pathogens, a review of the regulatory environment and safety control measures under which sheep milk is produced and Roquefort cheese manufactured, and verification of the implementation of control measures. The draft assessment report was ultimately published on March 23, 2005.[8] FSANZ found that all hazards that were considered potentially significant in Roquefort cheese were well managed through on-farm systems. Further, Roquefort was manufactured using application of Hazard Analysis and Critical Control Points (HACCP)–based controls during processing. These procedures, when applied in combination with standard operating procedures (SOPs) and good manufacturing practice (GMP) as determined and controlled by the Confederation of Roquefort Producers, allow the production of a microbiologically safe product. FSANZ acknowledged in the assessment that while normal food manufacture premise conditions are not applied, the controls and checks used during

Roquefort manufacture are adequate to ensure product safety. The risk assessment notes that the cheesemaking is "performed in technologically advanced premises staffed by suitably qualified and experienced people working to legislated and other commercial requirements." All plants have HACCP plans that are regularly updated. "Government monitoring by the Departmental Veterinary Services Directorates (DDSV) was found to be well implemented particularly in relation to animal health testing and records and the auditing of manufacturers."

In other words, the French system for regulation of raw milk safety and the subsequent manufacture of Roquefort cheese was deemed to be comprehensive and adequate. FSANZ noted the severity of sanctions against producers and manufacturers who fail to adhere to the requirements of the Ministerial Orders and the requirements of the Confederation of Roquefort Producers. Further, the regulatory system was determined to operate in a manner consistent with the Codex Code of Hygienic Practice for Milk and Milk Products. An on-site audit conducted by FSANZ verified the enforcement by the French government of control measures implemented by the Confederation of Roquefort Producers. The scientific evaluation of the safety of Roquefort cheese draft assessment report concluded that if Roquefort cheese is manufactured according to the submitted regulatory and industry processes, its consumption poses a low risk to public health and safety.

Based upon the results of the assessment, this traditional, blue-veined raw milk cheese is allowed to be imported into Australia. Parameters noted to contribute to Roquefort safety include a target acidity schedule and rapid acidification (to pH 4.8) of curd, a salt content of 3 percent, and 90 days of aging that promotes curd desiccation and moisture loss, leading to a final water activity of 0.92, where, combined with the other parameters, pathogens are unlikely to persist.

In conclusion, rigorous scientific studies and comprehensive risk assessments conducted in Italy, Switzerland, Canada, Australia, and elsewhere confirm the safety of cheeses made from raw milk. Why, then, does the FDA continue its quest to mandate pasteurization of all milk used for cheesemaking?

PART TWO

How Did We Get Here?

CHAPTER 5

EU versus US Regulations

The famous French general and statesman Charles de Gaulle once stated, "Only peril can bring the French together. One can't impose unity out of the blue on a country that has 265 different kinds of cheese." The same might be said for regulatory approaches to ensure cheese safety across the globe.

Cheese is a product that has long enjoyed global distribution, despite differences in regulatory approaches among the nations that import and export cheese. Following World War II, many countries agreed that international trade of food and other commodities should be better regulated to achieve open, fair, and undistorted competition. This resulted in the establishment of the General Agreement on Tariffs and Trade (GATT Agreement) in 1947, which created a system of rules accompanying free trade.[1] Unfortunately, some countries protected their local industries by imposing unjustified safety requirements, and GATT rules were inadequate to prevent this from happening. In response, the Agreement on the Application of Sanitary and Phytosanitary Measures (the SPS Agreement), was created in 1994 to address decades-old nontariff barriers.[2] Under SPS, countries possess the right to protect the health and life of consumers, animals, and plants against pests, diseases, and other threats to health. This must be balanced by preventing unjustified measures to unfairly impede international trade. The concept of *appropriate level of protection* (ALOP), also known as acceptable level of risk, became a term of art. ALOP is defined as the "level of protection that a country decides is appropriate to protect human, animal, or plant life within its territory." To avoid misuse of ALOP, countries must justify health risk with scientific risk assessment that is transparent, fair, and consistent. GATT was replaced by the World Trade Organization (WTO) in 1995, but GATT still exists as the WTO's umbrella treaty for trade in goods. The SPS Agreement importantly grants WTO members the right to challenge scientific findings of other countries to restrict importation of foods.

The WTO oversees the rules of trade among the 164 member nations that are responsible for 98 percent of world trade.[3] In 1961, the Food and Agriculture Organization (FAO) of the United Nations created the Codex Alimentarius Commission. Also known as the Food Code, the *Codex Alimentarius* is the global rule book for consumers, food producers and processors, national food control agencies, and international food trade. The purpose of the *Codex Alimentarius* is to protect the health of consumers, to ensure fair practices in the food trade, and to coordinate all work regarding food standards so as to facilitate global trade. In a global marketplace, it is obviously advantageous to have universally uniform food standards for consumer protection. WTO's SPS Agreement designates the joint FAO/WHO Codex Alimentarius Commission as the relevant standard-setting organization for food safety.[4] The WTO states that "to harmonize sanitary and phytosanitary measures on as wide a basis as possible, Members shall base their sanitary or phytosanitary measures on international standards, guidelines or recommendations."

However, in developing food standards, there is often disagreement among member countries of Codex. Such an example occurred at a 1997 meeting of the Codex Committee on Food Hygiene, when a draft outline was prepared by the United States, France, and the International Dairy Federation (IDF) on development of a milk code. The Codex Committee on Food Hygiene agreed to stop work on the "Proposed Draft Code of Hygienic Practice for the manufacture of unripened cheese and ripened soft cheese."[5] The United States proposed: "Pasteurization or an equivalent measure approved by the official agency having jurisdiction shall be used to achieve the ALOP." France objected, noting, "Common hygiene provisions provide adequate health protection without pasteurization."

The US position was that raw milk and raw milk products are potentially hazardous foods that support the growth of pathogens such as *Listeria*, *Salmonella*, *E. coli*, and others. The United States stated that cheese poses a particularly high health risk because it is usually ready-to-eat and will not be cooked before consumption, but noted that scientifically accepted processes control microbiological safety threats. These processes can include pasteurization, heat treatment, sterilization of milk; aging of cheese; and new technologies that had not yet been developed. Meeting delegates noted that this position differed from the EU Commission's position that consumer safety is best protected when strict veterinary and sanitary practices are followed from production to consumption, including:

- Using raw milk from herds in good health with regular veterinary inspections subject to sanitary controls
- Using milk that is collected, transported, and transformed into cheese within a short period of time, applying strict hygiene
- Educating consumers about the proper storage and shelf life of cheese

The French delegation took the position that common hygiene provisions provide adequate protection without mandatory pasteurization. Products such as raw milk cheese and traditionally cured meats are prized by European consumers, and their safety is assumed based on centuries of experience. The French contend that most Americans view traditional practices as unsafe, and that our food safety policies rely instead on scientific views and provide a climate that encourages the use of new technologies.

As stated by Hornsby:

> The case of raw-milk cheese also suggests that the federal government maintains a great deal of power as it is responsible for international negotiations and can advance its preferred policy position at an international standard setting organization. Such a power needs to be utilized carefully and with sensitivity to political dynamics that exist . . . especially in international negotiations over matters pertaining to an area like food safety, where competency is shared and perceptions of risk, science, economic imperatives, and culture intersect.[6]

Put another way, our US trade representatives should ensure that their opinions and expressed views are representative of *all* US food manufacturers and not a chosen few who, using their lobbying power, drown out the voices of small-scale producers or consumers who desire choice in their foods.

The quality of raw milk used for cheesemaking is perhaps the most important consideration when producing raw milk cheeses. European cheesemakers pay great attention to raw milk quality, knowing that this will in turn dictate cheese quality. Raw milk quality is assessed through an analysis of bacterial numbers and somatic cell counts. When low, these indicators generally indicate high-quality milk, but as numbers of bacteria and somatic cells increase, the potential for contamination of milk and cheese with pathogens also increases. Most artisan cheesemakers have a program in place to regularly monitor and control levels of bacteria and somatic cell counts in milk. In addition to potential safety concerns, as raw milk bacteria and somatic cell counts increase, cheese quality and cheese yield may decrease. Raw milk quality decreases during refrigerated storage; therefore, using milk as soon as possible for cheesemaking protects cheese quality. In artisan cheesemaking, if milk is produced on the farm, the reduced time from milking to cheesemaking ensures the manufacture of high-quality cheese. Immediate manufacture of milk into cheese without cooling reduces the opportunity for the growth of undesirable bacteria. Conversely, when milk is cooled and held in a bulk tank, the potential for the growth of cold-loving psychrotrophic pathogens and other bacteria is increased. A most interesting publication from France appeared in the September 2004 edition of *Applied and Environmental Microbiology*, a publication of the prestigious American Society for Microbiology.[7] In this article, the authors state that the diversity of raw milk microflora is the greatest contributor to the taste and flavor of raw milk cheeses. In contrast to widely used US dairy practices, these authors cite work that has shown that intensive washing of equipment used to store milk under refrigeration, along with the use of udder preparations, leads to high levels of *Pseudomonas* contamination, a leading spoilage organism of milk. Refrigeration of milk for 24 hours at 39°F (4°C) led to the emergence of populations of *Listeria* and *Aeromonas hydrophila*, two psychrotrophic pathogens, as well as spoilage flora. All of these organisms are negative contributors to milk and cheese quality. In contrast, in milk collected directly from the cow that did not undergo cooling, *Lactococcus lactis*, an important member of the lactic acid bacteria that is often used as a starter culture to facilitate lactic acid production, was confirmed to be the major raw milk species.

The European Community Directives 92/46 and 92/47 contain regulations for the production and marketing of raw milk, heat-treated milk, and milk-based products. These regulations establish hygienic standards for raw milk collection and transportation that focus on issues such as temperature, sanitation, and microbiological standards. These standards enable production of raw milk of the highest possible quality. Some regulations, such as those of the EU, have established microbiological limits at the sell-by date for products such as cheeses. With respect to regulations that govern the use of raw milk for cheesemaking, limits have been established for *Staphylococcus aureus* in raw milk. Finished cheeses must meet specific hygienic standards, in which case the presence of *S. aureus* and *E. coli* indicate poor hygiene. US cheesemakers might benefit from the adoption of some of the EU directives, as will be discussed.

To the points raised by France and the EU at the 1997 meeting of the Codex Committee on Food Hygiene, my laboratory conducted research through a grant from USDA to look at the microbiological quality of raw milk that was specifically used for artisan cheese production. In studies conducted during the summer of 2006, we examined the microbiological quality of raw milk used for cheesemaking at farmstead cheese operations in Vermont. We sampled milk throughout the summer, during the time when microbiological risk is highest because of warm summer temperatures. The results of our investigation revealed that, for most (70 percent) of the milk we sampled, the raw milk exceeded the microbiological standards expected for pasteurized milk. Further, we found a low incidence of pathogens of concern in raw milk (*Salmonella, Listeria, E. coli* O157:H7, and *Staphylococcus aureus*).[8]

Similar results were obtained from a follow-on study that we conducted in 2008.[9] The most prevalent pathogen found in milk samples was *Staphylococcus aureus* (Staph). This is the target organism used by the EC to assess the safety of raw milk cheese. The organism produces a toxin that can cause foodborne illness when numbers increase to high levels. For cheeses made from pasteurized milk, the EC targets Staph levels as well as nontoxigenic *E. coli*. *E. coli* is a hygienic indicator, not a safety indicator. It is not heat-resistant, and it will not survive pasteurization or other heat treatment of milk. Therefore, if a pasteurized milk cheese contains high levels of *E. coli*, the organisms could only be present as a result of contamination during processing, after pasteurization. To ensure cheese hygiene, Regulation (EC)

2073, established in 2005, permits no more than 1,000 *E. coli* per gram of cheese made from heat-treated milk.

In establishing microbiological criteria for foods, the EU states that criteria should enhance food safety, be feasible in practice, and be based on scientific risk assessment. Scientific risk assessments have shown that Staph presents a significant food safety risk in cheeses made from raw milk. To control this hazard, the EU requires cheesemakers to test cheeses at the time during the cheese manufacturing process when the number of staphylococci are expected to be the highest. If test results indicate a problem, producers are asked to make improvements in hygiene and the selection of raw materials. If numbers are sufficiently high where the presence of toxin is suspected, the cheese batch must be tested for staphylococcal exterotoxins.

Over the last 20 years, there have been challenges to the 60-day aging rule, made particularly by the FDA. Research conducted in the 1990s showed that *Salmonella typhimurium*, *E. coli* O157:H7, and *Listeria monocytogenes* can survive beyond the mandatory 60-day aging period. In a referral to the National Advisory Committee on the Microbiological Criteria for Food (NACMCF) in 1997, the FDA asked the NACMCF if a policy revision was necessary, as 60 days of aging may not be sufficient to provide adequate public health protection. The FDA noted that independent qualifying bodies in other countries, such as the Institute of Food Science and Technology (IFST) in the UK, adopted a position statement noting the potential health hazards due to pathogens in raw milk cheese. The NACMCF concluded that evidence in the scientific literature provided support that pathogens could survive the 60-day aging process for cheeses manufactured from unpasteurized milk.

It is clear that our approach in the United States toward raw milk cheesemaking stands in contrast to approaches and scientific assessments conducted in other parts of the globe. Are there forces at work that explain why our rules, regulations, and approaches stand in stark contrast to regulations elsewhere that have been shown to protect public health and allow consumer access to high-quality traditional cheeses? As a bona fide cheese culture develops in the United States, would we not be wise to adopt harmonized testing procedures to assure cheese safety? And if not, why not?

CHAPTER 6

Redefining Pasteurization

What is pasteurization? In general terms, it is defined as the application of mild heat to foods to destroy microorganisms. Historically, pasteurization has been used for two purposes: to protect public health by destroying pathogenic microorganisms, and to improve product quality by destroying spoilage microorganisms. Louis Pasteur (1822–1895), the eminent French scientist, pioneered the concept of pasteurization. In the 1860s, he discovered that grape juice could be mildly heated to eliminate spoilage organisms without adversely affecting the quality of wine. This concept of pasteurization was later used for public health protection. Pasteurization of milk was first proposed in 1886 by the German scientist Franz Ritter von Soxhlet to improve the safety of milk fed to infants.

Pasteurization of milk and dairy products has no doubt improved public health worldwide. Milk-borne diseases such as tuberculosis, diphtheria, scarlet fever, typhoid fever, Q fever, brucellosis, anthrax, and foot-and-mouth disease were rampant prior to milk pasteurization. Between the years 1900 and 1939, typhoid fever caused by *Salmonella typhi*, scarlet fever from *Streptococcus pyogenes*, and diphtheria caused by *Corynebacterium diphtheriae* were the leading causes of milk-borne disease in the United States. Between the years 1912 and 1937, some 65,000 deaths from tuberculosis occurred in England and Wales from milk consumption. *Mycobacterium tuberculosis* was the original target for milk pasteurization, when minimum milk holding times of 30 minutes at temperatures of 142°F (61.1°C) were established. These holding times and temperatures were later increased (30 minutes at 145°F/63°C for vat pasteurization; 161°F/72°C for 15 seconds for high-temperature short-time pasteurization (also known as HTST) to ensure destruction of *Coxiella burnetii*, the causative agent of Q fever.

It is ironic that, as the United States was holding firm on its position at the 1997 meeting of the Codex Committee on Food Hygiene that "pasteurization or an equivalent process" was required to achieve dairy product safety, a funny thing was happening back in the United States. Pasteurization was being redefined by Congress. To this day, most Americans are unaware of this change. Why was this happening? Due to outbreaks of illness linked to ground beef, commercial development of an irradiation process to ensure ground beef safety was being undertaken. On May 16, 2000, irradiated frozen ground beef patties first became available to US consumers. However, proponents of irradiation were concerned about the lack of consumer acceptance of the use of irradiation as a technology to safely process foods. Irradiated foods must bear the radura symbol, which many consumers interpreted as a warning. By redefining the labeling term *pasteurized* to include food products that had been irradiated, irradiation proponents hoped there would be better acceptance of irradiated foods. Senator Tom Harkin of Iowa had sponsored this legislative change, and it was signed into law on May 13, 2002.

The main intent of the legislation, named the Farm Security and Rural Investment Act (FSRIA), popularly known as the Farm Bill, was to provide for the continuation of a variety of agricultural programs such as commodity programs, conservation, trade, credit, and rural development. Redefining pasteurization was included under a section of the law that dealt with misbranding.

Title X, subtitle I, Section 10808(b) of FSRIA broadened the definition of pasteurization by mandating that Section 403(h) (Misbranded Food) of the Federal Food, Drug, and Cosmetic Act (FFDCA) be amended to include a definition for pasteurization. Section 403(h) of the FSRIA amendment discusses misbranding of foods, and it includes pasteurization in the discussion as follows:

> Sec. 403 [343] A food shall be deemed to be misbranded—
> (h) If it purports to be or is represented as—

(3) A food that is pasteurized unless—

(A) such food has been subjected to a safe process or treatment that is prescribed as pasteurization for such food in a regulation promulgated under this Act; or

(B)(i) such food has been subjected to a safe process or treatment that—

(I) is reasonably certain to achieve destruction or elimination in the food of the most resistant microorganism of public health significance that are likely to occur in the food

(II) is at least as protective of the public health as a process or treatment described in subparagraph (A)

(III) is effective for a period that is at least as long as the shelf life of the food when stored under normal and moderate abuse conditions; and

(IV) is the subject of a notification to the Secretary, including effectiveness data regarding the process or treatment[1]

As explained previously, milk is the product to which pasteurization has been most commonly applied. Milk is subjected to mild heat at either 145°F/63°C for 30 minutes (vat pasteurization), or 161°F/72°C for 15 seconds (high-temperature short-time pasteurization). Both of these time/temperature parameters result in the destruction of *Coxiella burnetii*, the most heat-resistant pathogen of concern in milk. Since the pasteurization process targets the most heat-resistant pathogen of concern, other, less heat-resistant pathogens such as *Salmonella* and *Listeria* would also be destroyed. The FDA sought the advice of its National Advisory Committee on Microbiological Criteria for Foods (NACMCF) to determine what would be required to establish the equivalence of alternate methods of pasteurization. A number of questions were considered, including:

- What are the scientific criteria that should be used to determine if a process is equivalent to pasteurization?
- What, if any, further research is needed to determine criteria?
- What is the most resistant microorganism of public health significance for each process?

- What data need to be acquired to scientifically validate and verify the adequacy of a proposed technology?
- How much data would be considered adequate?
- To what degree can models and published literature be relied upon as contributing to validation?
- What biological hazards might be created as a consequence of the pasteurization treatment?

Ironically, one of the processes considered to be potentially equivalent to pasteurization was fermentation, which is exactly what cheesemaking is based upon. NACMCF's working definition of pasteurization became:

> Any process, treatment, or combination thereof, that is applied to food to reduce the most resistant microorganism(s) of public health significance to a level that is not likely to present a public health risk under normal conditions of distribution and storage.[2]

Cheesemaking and cheese aging are described quite nicely by this definition. This definition also shows why the concept of equivalence as applied to cheesemaking, and as used by Food Standards Australia New Zealand (FSANZ), is a scientifically valid approach to achieve safety. The NACMCF definition aligns with the *Codex Alimentarius* Code of Hygienic Practice for Milk and Milk Products, which states:

> From raw material production to the point of consumption, dairy products produced under this Code should be subject to a combination of control measures, and these control measures should be shown to achieve the appropriate level of public health protection.[3]

When the FDA pushes for mandatory pasteurization of all milk intended for cheesemaking, it fails to acknowledge that the legal definition of pasteurization was broadened by Congress. Further, as we have seen, Australia established the equivalence of alternative methods of pasteurization when assessing the safety of hard Italian and Swiss cooked-curd cheeses, as well as Roquefort cheese. There is a strong body of scientific evidence supporting cheese safety. Why is this important evidence disregarded?

CHAPTER 7

FDA's Assault on Artisan Cheese, Phase 1

The Domestic and Imported Cheese Compliance Program

In November 1998, the FDA initiated the Domestic and Imported Cheese and Cheese Products Compliance Program. Why at this time? As with the FDA's request in 1997 to the National Advisory Committee on Microbiological Criteria for Foods (NACMCF) to revise the 60-day aging rule, the timing of the Domestic and Imported Cheese and Cheese Products Compliance Program is significant and will be discussed later in the book. The objectives of this program were for the FDA to conduct inspections of domestic cheese firms and to examine samples of domestic and imported cheese for microbiological contamination. The target pathogens were *Listeria monocytogenes*, *Salmonella*, *E. coli* (and, if numbers of *E. coli* were greater than 10,000 cfu/g, testing for enteropathogenic *E. coli* would be conducted); *E. coli* O157:H7, and *Staphylococcus aureus*. Despite the large volume of testing and inspection conducted to date, the FDA has not publicly shared the results of the Domestic and Imported Cheese and Cheese Products Compliance Program. On June 14, 2004, a Freedom of Information Act (FOIA) request was submitted to the FDA on behalf of the Cheese of Choice Coalition (CCC). The FOIA request was for all tests obtained under the agency's Domestic and Imported Cheese and Cheese Products Compliance Program from June 1998 to June 2004. On August 7, 2007, Richard Koby, the lawyer representing the CCC, agreed, at the FDA's urging, to amend the time period of the data request to January 1, 2004, through December 31, 2006. On February 7, 2008, almost four years after the initial FOIA request, the FDA finally sent to the CCC the requested data.

My lab analyzed and shared results of the analysis with CCC members, and ultimately published the results of this analysis in 2011.[1] During the years 2004 through 2006, the FDA tested 17,324 domestically produced as well as imported cheese samples. In support of the EU's approach to ensuring cheese safety, *S. aureus* was the pathogen most commonly detected by the FDA, found in 6.9 percent of 3,449 cheeses samples tested. A total of 3,520 cheese samples were analyzed for the presence of *Salmonella*, with 1.3 percent testing positive. In general, the majority of positive samples came from soft or soft-ripened cheeses produced in Mexico or Central America. Throughout the same period, the FDA examined 2,181 cheese samples for the presence of *Listeria monocytogenes*. A total of 52 samples (2.4 percent) were positive for this pathogen. More than half (52 percent) of the positive samples were from Mexican-style soft cheese (17 positive samples) or soft-ripened cheese (10 positive samples). When data from all years are combined, the overall incidence of *L. monocytogenes* in domestic cheese samples (1.2 percent) was lower than that in imported products (3.2 percent). The incidence of *L. monocytogenes* in imported samples decreased from 4.25 percent in 2004 to less than 1 percent by 2006, suggesting increased control and/or regulatory compliance by cheese producers abroad.

Curiously, of 3,360 cheese samples tested by FDA during the three years between 2004 and 2006 for *E. coli* O157:H7, only three cheese samples were positive. Two cheese samples tested in 2004 were positive, consisting of one soft-ripened cheese sample from Honduras and one sample of Mexican-style soft cheese from Mexico. In 2005, one sample of Mexican-style soft cheese tested positive, and no samples tested positive in 2006. If you recall back to chapter 2, *E. coli* O157:H7 was the pathogen that FDA had used in experiments with Cheddar cheese, and its survival beyond 60 days of aging continues to be used as the scientific basis for proposing changes to the 60-day aging rule. Had *E. coli* O157:H7 presented a severe public health threat in cheese, one would have expected a much higher incidence of this dangerous pathogen in the sampled cheese products. FDA's Domestic and Imported Cheese and Cheese Products Compliance Program data therefore supported the premise that the 60-day aging rule was effectively working to protect public health.

FDA Establishes Stringent *E. coli* Criteria

Despite the findings from the Domestic and Imported Cheese and Cheese Products Compliance Program concerning the low incidence of *E. coli* O157:H7 in cheese, it was therefore surprising when, in 2009, the FDA published a draft Compliance Policy Guide (CPG 7106.08) Sec. 527.300 Dairy Products—Microbial Contaminants and Alkaline Phosphatase Activity. Compliance Policy Guides (CPGs) provide direction to FDA investigators and the regulated industry as to how FDA will enforce food safety law. This 2009 draft CPG stated:

> The presence of *Escherichia coli* in a cheese and cheese products made from raw milk at a level greater than 100 MPN/g (Most Probable Number per gram) indicates insanitary conditions relating to contact with fecal matter including poor employee hygiene practices, improperly sanitized utensils and equipment, or contaminated raw materials.[2]

The August 1, 1986, Compliance Policy Guide 7106.08 (which was revised in 1996) had established an *E. coli* limit of 10,000 colony forming units (cfu) per gram of cheese. The Domestic and Imported Cheese and Cheese Products Compliance Program guidelines established in 1998 took regulatory action on cheeses testing above this limit, requiring testing for enteropathogenic *E. coli* only in cheeses above the 10,000 cfu/g *E. coli* limit. EC 2073/2005 guidelines permitted no more than 1,000 *E. coli* in cheese made from heat-treated milk, but no criteria were established by the EC for cheeses made from raw milk. The FDA's CPG changes proposed for *E. coli* were a drastic departure from previous FDA guidance as well as EC regulations, yet the FDA offered no rationale for these significant changes.

Unbeknownst to the artisan cheese community, the FDA had requested, in the Federal Register, public comments on this draft guidance and received a comment from the American Dairy Products Institute, which stated: "In our view, the permissible level of *Escherichia coli* should be set according to standards of food safety without regard to the treatment of the milk itself. Stated another way, the guidance should be set at a uniform level to ensure food safety across all covered dairy products." Despite this recommendation being in total conflict with EC criteria for cheese and with guidance from other respected scientific bodies around the globe, in 2010, the FDA reissued its Compliance Policy Guide, and under part B, Nontoxigenic *Escherichia coli*, it states:

> Dairy products may be considered adulterated within the meaning of section 402(a)(4) of the Act (21 U.S.C. 342(a)(4)), in that they have been prepared, packed, or held under insanitary conditions whereby they may have become contaminated with filth when *E. coli* is found at levels greater than 10 MPN per gram in two or more subsamples or greater than 100 MPN per gram in one or more subsamples.[3]

Curiously, the microbiological standard for pasteurized milk is that it cannot contain greater than 10 coliform bacteria per milliliter of milk. Was the FDA forcing a pasteurization standard on all cheese? Had FDA changed these standards at the request of lobbyists working on behalf of US industrial cheese producers? Was the FDA rewriting cheese standards to conform to requirements that the US industrial cheese producers wanted to see in place? And if so, why?

As previously mentioned, the American Cheese Society (ACS) is the organization that represents the artisan cheesemaking community in the United States. ACS members were caught off guard by the 2009 draft and the 2010 final CPG changes, and many cheesemakers became aware of these new rules only when they received unannounced inspections and were told that the FDA had found evidence of filth in their products. When the ACS questioned the FDA as to why non-toxigenic *E. coli* in raw milk cheese was being targeted, the FDA responded in an August 29, 2014, letter addressed to ACS president Peggy Smith and ACS executive director Nora Weiser. William A. Correll, Jr., director of the office of compliance at the FDA's Center for Food Safety and Applied Nutrition, explained that representatives of the FDA had met with the ACS Board of Directors in Sacramento during the ACS conference held during July 29 to July 31, 2014. During that meeting, the ACS had asked the FDA to consider revising the non-toxigenic *E. coli* standard it had established in the 2010 CPG. Among other issues, the ACS expressed concern about the impact the non-toxigenic *E. coli* standard was having on the commercial flow of French Roquefort cheese. The FDA told the ACS that it had initiated a pilot sampling program involving 1,600 cheese samples, "aimed at aligning with the goals of the Food Safety Modernization Act (FSMA), which mandates a risk-informed and preventive approach to food safety."

The FDA further wrote to the ACS:

> Through this pilot, we are learning more about how often select foods, such as 60-day aged raw milk cheese, become contaminated with foodborne pathogens, and what patterns, if any, may help predict potential contamination in the future.

The FDA informed the ACS that, to date, it had found a pathogen contamination rate of less than 1 percent. While the FDA indicated it had adjusted its sampling due to misapplication of criteria in the sampling plan, it was not rolling back its non-toxigenic *E. coli* standards established in the 2010 CPG, but noted:

> With the adjustments to testing and regulatory action criteria discussed previously, approximately 95% of the raw milk cheese examined thus far meet the n=5, c=2, m=10 MPN/g, M=100 MPN/g criteria.

This kerfuffle also caused the FDA on September 8, 2014, to issue a constituent update:

> Recent media reports have incorrectly indicated that the FDA is banning Roquefort and other cheeses.
>
> Earlier in 2014, nine producers of Roquefort, Tomme de Sovie, Morbier, and other cheeses tested above threshold levels set in 2010 for a particular type of bacteria called non-toxigenic

> *E. coli.* While these bacteria don't cause illness, their presence suggests that the cheese was produced in unsanitary conditions.
>
> The FDA has been working with the American Cheese Society (ACS) to learn more about artisanal cheeses and measures that cheesemakers take to ensure their products are safe. After hearing ACS' concerns about the test results, the FDA adjusted its criteria for taking regulatory action based on them. As a result, 95 percent of the cheese sampled tested below the level at which FDA would take regulatory action, and six of the nine cheese producers placed on Import Alert 12-10 for exceeding bacterial counts have been removed from that list and can resume sales and distribution in the U.S.
>
> The FDA remains dedicated to ensuring a safe and wholesome food supply using the latest science to protect human health, and promoting dialogue with industry, consumers and other interested parties. The FDA is committed to working and sharing an open dialogue with the artisanal cheesemaking community. Of course, we welcome input from the public at any time and we continue to meet and share information with the artisanal cheesemaking community on this and other topics.

The August 29, 2014, letter from the FDA to the ACS highlights a number of issues of concern to both cheesemakers and scientists. Although the letter references the FSMA and a sampling assignment that was under way, the 2009 draft CPG changes predated the FSMA, as the FSMA was not signed into law until 2011. So 2009 CPG changes had nothing to do with the FSMA.

Further, the EC had established microbiological criteria for cheese in 2005, a fact not noted by the FDA. As 2009 CPG criteria were not consistent with 2005 EC criteria, it was clearly foreseeable that EU-produced cheeses, particularly those made from unpasteurized milk, would not meet established 2009 CPG draft criteria, nor 2010 final CPG criteria. They had been produced to a different regulatory standard in use in the EU.

As I wrote to FDA's Deputy Commissioner Michael R. Taylor on December 8, 2014, our European colleagues established microbiological criteria for cheese in 2005 through EC Regulation No. 2073/2005 (see

Table 7.1. Regulation (EC) 2073/2005, microbiological criteria

Food	Bacteria	Sampling Plan		Limits (cfu/g)		Method
		n	c	m	M	
With thermal treatment	*E. coli* (hygienic index)	5	2	10^2	10^3	ISO 16649-1
	Staphylococcus positive coagulase	5	2	10^2	10^3	ISO 6888-2
Raw milk cheese	*Staphylococcus* positive coagulase	5	2	10^4	10^5	ISO 6888-2

Note: n = sample units tested; c = max no. samples between m and M
M = quality/safety limit; m = GMP limit

table 7.1). For cheeses made from milk receiving heat treatment (pasteurization or otherwise), microbiological limits are established for *E. coli* (hygienic index) and for coagulase-positive *Staphylococcus* (food safety index) where out of five tested samples, no more than two can contain Staph or *E. coli* at levels between 100 and 1,000 cfu/g, and no sample can contain more than 1,000 cfu/g *E. coli* or Staph. For cheeses made from raw milk, only microbiological criteria for coagulase-positive *Staphylococcus* are established (n = 5, c = 2, m = 10^4, M = 10^5). For both raw milk cheese as well as cheese made from milk receiving heat treatment, samples must be taken at the time during the manufacturing process when the number of staphylococci is expected to be the highest. It is critical to note that no limits are established for *E. coli* in raw milk cheese. In cheeses made from pasteurized milk, because *E. coli* are not heat tolerant, the presence of *E. coli* indicates post-process recontamination and thus provides a meaningful indicator of hygiene. In cheeses made from raw milk, presence of *E. coli* does not provide a scientifically meaningful index of sanitary conditions or filth.

On September 8, 2014, Q13 Fox in Seattle reported a story, "'Confusion' over FDA rules could endanger your favorite French cheese." The Q13 news staff wrote, "*The Los Angeles Times* reports that Food and Drug Administration rules could effectively ban age-old recipes for cheeses like Roquefort, St. Nectaire, Morbier and Tomme de Savoie" and that cheese retailers were having trouble obtaining imported French cheese. Peter McNamara, the manager of the Righteous Cheese Shop, was quoted as saying that imported soft cheeses of any kind were hard to find. "There's a lot of confusion (over the rules) . . . getting hold of any soft cheese is really a pain," McNamara said.

Q13 reported that the FDA was not aware of a cheese shortage. When asked about *E. coli* standards, FDA's Lauren Sucher stated, "The current level is in line with standards around the world, and FDA expects that properly manufactured milk products, whether made from raw milk or pasteurized milk, should not be affected."[4] Of course, not only was this statement misleading, it was also factually incorrect. On September 10, 2014, the ACS posed follow-up questions to the FDA:

> We are being asked why the 2009 Draft was initially created, and ultimately, why the changes from that draft to the final 2010 CPG were made. In particular: Why were the proposed permissible levels for non-toxigenic *E. coli* changed, and the explanatory language regarding raw milk cheeses removed, from the 2009 CPG draft to the 2010 final CPG? We have also received a number of inquiries regarding the scientific data and/or correlation between the presence of non-toxigenic *E. coli* and the incidence of pathogens in cheese overall, and how that data supports the determination of allowable limits.

On October 30, 2014, the ACS received a nine-page letter from the FDA's Dr. Nega Beru and William Correll with answers to the questions it had posed. The responses are very interesting.

In response to why the FDA changed the 1996 CPG, the FDA responded:

> Following issuance of the 1996 CPG, the domestic dairy industry shared its concerns with FDA regarding the permissible level of non-toxigenic *E. coli*. The concerns were that permitting up to 10,000 cfu/gm of product 1) creates the appearance that the U.S. allows some domestically manufactured dairy products

> to be produced under insanitary conditions, and 2) poses an obstacle to exporting domestically manufactured dairy products, as export markets question why U.S. dairy products would be permitted to have such levels of non-toxigenic *E. coli.*
>
> After reviewing these concerns, FDA in 2009 initiated a review of the scientific literature to determine if there was a scientific basis for changing the 10,000 cfu/g nontoxigenic *E. coli* criterion, and if so, to what level. Based on its review, FDA determined that a change to the 1996 CPG non-toxigenic *E. coli* criteria was supported by the literature and warranted.

On page two, the FDA indicated:

> FDA's use of non-toxigenic *E. coli* as an indicator organism is longstanding and consistent with the advice of the ICMSF (Volume 7, Chapter 5, and section 5.7.2.1 "Indicators") and other literature (Smoot and Pierson (1997) and Chye et al. (2004)).

As stated previously, FDA guidance contradicts ICMSF (International Commission on Microbiological Specifications for Foods) guidance, which specifically addresses *E. coli* in cheese (volume 2, chapter 15, pages 168–69).[5] The FDA further stated that "the presence of non-toxigenic *E. coli* is widely considered an indication of fecal coliform bacteria contamination, potential insanitation, and the possibility that Gram negative enteric pathogens (e.g. Enterotoxigenic *E. coli* and *Salmonella* spp.) might be present." For the reasons expressed by ICMSF along with other data, this statement is not scientifically accurate. As direct testing methodology is available for pathogens, there is no scientific reason to test raw milk or raw milk cheese for non-toxigenic *E. coli*. Many raw milk cheesemakers test finished products directly for *Salmonella* and enterohemorrhagic *E. coli*, and such an approach should directly address the stated FDA safety concerns.

Further, the FDA stated, "Excessive levels of non-toxigenic *E. coli* in a food, including cheese, is an indication to FDA that the food was prepared, packed, or held under insanitary conditions whereby it may have become contaminated with filth, or whereby it may have been rendered injurious to health." The term *excessive* offers no guidance to cheesemakers. Using the FDA's current guidance of "not more than 100 MPN/g non-toxigenic *E. coli*" is inconsistent with guidance established in the EU and is not aligned with modern preventive, risk-based approaches to food safety. Extensive research on artisan cheese conducted over the last 10 years in my laboratory and elsewhere has shown that the establishment of monitoring targets for *Salmonella*, *L. monocytogenes*, and *S. aureus* is more appropriate to provide assurance of cheese safety. What benefit will *E. coli* testing provide over and above limits for the aforementioned pathogens?

When ACS asked why the FDA's final CPG published in December 2010 eliminated the two different criteria for non-toxigenic *E. coli* (one for raw milk cheese and one for pasteurized milk cheese), eliminated language in the draft 2009 CPG with respect to low levels of non-toxigenic *E. coli* in raw milk products or products made from raw milk, and adopted a uniform criterion for non-toxigenic *E. coli* in cheese made from either raw or pasteurized milk, the FDA responded:

> FDA received comments on the draft 2009 CPG from four stakeholders. Among the comments received, the International Dairy Foods Association (IDFA) questioned the scientific justification for FDA's different criteria for considering regulatory action in raw milk cheese and pasteurized milk cheese with regard to non-toxigenic *E. coli*. The American Dairy Products Institute (ADPI) likewise commented that a uniform level should be set for non-toxigenic *E. coli* as follows: "ADPI understands that the distinction between products made from raw milk or pasteurized milk is based on the inevitable presence of this contaminant in raw milk and raw milk products, even when GMPs are followed. Nevertheless, in our view, the permissible level of *Escherichia coli* should be set according to standards of food safety without regard to the treatment of milk itself. Stated another way, the guidance should be set at a uniform level to ensure food safety across all covered dairy products."

The FDA went on to further explain:

> In light of those comments, the agency re-examined the scientific studies it had relied on to develop the criteria in the draft 2009 CPG and concluded that the same criteria for non-toxigenic *E. coli* could be applied successfully to both raw milk and pasteurized milk cheese, FDA modified the microbial criteria for considering possible regulatory action when finding non-toxigenic *E. coli* in cheese made from raw milk from levels between 100 MPN/g and 1000 MPN/g in a certain number of subsamples tested (draft 2009 CPG) to levels between 10 MPN/g and 100 MPN/g in the same number of subsamples (final 2010 CPG); and from not more than 1000 MPN/g in one or more substamples (draft 2009 CPG) to not more than 100 MPN/g in one or more subsamples (final 2010 CPG).

Where is the scientific evidence to support these statements? To what extent did the FDA consult with international organizations and producers who import cheese to this country?

The FDA continued:

> In deciding upon a final level for M, FDA considered ICMSF advice that, as a general hygiene indicator, M should represent clearly unacceptable conditions of hygiene. The scientific literature, international standards in use, and FDA's own analytical results for non-toxigenic *E. coli* in cheese, led the agency to conclude that M at 100 cfu/g is consistently attainable and that exceeding this level in cheese is indicative of conditions meeting the adulteration standard of section 402(a)(4) of the FD&C act.

There are many problems with the FDA statements in these letters. The first is that it did not follow the ICMSF guidance. The ICMSF is recognized as the global leading scientific body for the establishment of microbiological criteria in foods. The ICMSF also provides global guidance for sampling plans for foods. In ICMSF Book 8, published in 2011, Table 23.7 shows the

ICMSF-established microbiological criteria for raw milk cheeses targeting *S. aureus*. An *E. coli* criterion was established only for cheeses made from pasteurized or heat-treated milk. In cheeses made from pasteurized milk, *E. coli* limits are established under a sampling plan where n = 5, c = 3, m = 10, and M = 10^2. Raw milk cheese is tested for *S. aureus* only, consistent with EC recommended sampling criteria. It is notable that for EC microbiological criteria for cheese, no limits were established for *E. coli* in raw milk cheese.

The FDA's position is also in conflict with prior guidance from the ICMSF. In 1986, the ICMSF wrote:

> While the coliform problem in cheese is well known, presence of these organisms in many cheese varieties is extremely difficult to prevent completely. With some varieties, if coliforms are present initially, it is virtually impossible to prevent their growth during manufacture or during the ripening period. In several types of cheese, *E. coli* can even be considered characteristic. With the exception of some strains of *E. coli*, high populations of coliforms are unlikely to present a health hazard. There is ample evidence that if pathogenic strains of *E. coli* (PEC) are present early in the cheesemaking process their numbers may increase to hazardous levels. However, in view of the scarcity of evidence of recurring outbreaks due to PEC in cheese and the high cost of routine testing, it is doubtful that establishment of end-product criteria for either coliforms or *E. coli* would be justified. Accordingly, no sampling plan is proposed.

Translated, this ICMSF statement indicates that cheese made from raw milk should not be tested for *E. coli*.

The FDA also wrote:

> Since FDA's scientific literature reviews in 2009 and 2010, researchers have published results of a survey of retail raw milk cheese for microbiological quality and the prevalence of foodborne pathogens. Among other parameters, the researchers examined retail raw milk cheese samples for the presence of non-toxigenic *E. coli*.

> Some 41 cheese samples were obtained nationwide from retail specialty shops, farmers markets, and online sources. The samples included Cheddar, Blue, Gruyere, Gouda, Romano, Monterey Jack, soft, semi-soft, semi-hard and hard cheese made from cow, goat and sheep raw milk.
>
> Only two samples of the 41 tested contained levels of nontoxigenic *E. coli* at or exceeding 10 cfu/g.

Correll and Beru said:

> The two positive samples included a Romano cheese made from sheep milk (30 cfu/g) and a cow's milk semi-soft cheese (10 cfu/g); both originated from the Northeast US. Ninety-five percent of the cheese tested in this study had non-toxigenic *E. coli* levels below 10 cfu/g. . . . Another study, O'Brien et al. (2009) also examined cheese for non-toxigenic *E. coli*, tangential to studying the occurrence of foodborne pathogens in Irish farmhouse cheese, and the researchers found that 79 percent of the raw milk cheese contained less than 10 cfu/g.
>
> Both domestically and foreign produced raw milk cheeses are meeting the three-class attribute sampling plan at a rate of approximately 95 percent under FDA's ongoing microbiological surveillance sampling assignment for raw milk cheese.

But the FDA neglected to indicate that it had been collecting *E. coli* data from the Domestic and Imported Cheese Compliance Program since 1998. Had it analyzed even some of these data, such as the data on 17,000 cheese samples supplied to the Cheese of Choice Coalition in 2008, it would have found that most of the cheese samples it examined could not meet the FDA's 2009 or 2010 CPG criteria. My laboratory conducted a retrospective analysis of data from the FDA's Domestic and Imported Cheese and Cheese Products Compliance Program results. Our analysis revealed that approximately 70 percent of the cheese samples that the FDA tested between January 1, 2004, and December 31, 2006, exceeded 100 MPN/g for *E. coli*. Since the FDA had been testing cheese samples since 1998, when it initiated the Domestic and Imported Cheese and Cheese Products Compliance Program, the FDA could have examined its own data and realized that the majority of tested cheese samples could not have complied with the 2010 CPG criteria. Or perhaps it did review the data, and is this why these stringent standards were established?

Anecdotal reports from cheesemakers at the ACS meetings in 2015 expressed the economic impact of the FDA's actions. Neal's Yard Dairy, an artisanal cheese retailer in London, had 25 percent of its shipments to the United States rejected in August 2015 alone.

Fortunately, US cheesemakers reached out to their members of Congress, who intervened to provide oversight of the FDA's actions. The issue was the subject of a bicameral letter drafted on December 3, 2015, that was sent to FDA Deputy Commissioner Michael R. Taylor. The letter was authored by Congressman Peter Welch of Vermont and 15 other members of Congress along with Vermont Senators Patrick Leahy and Bernie Sanders, and seven other members of the US Senate. The letter detailed the impact the new *E. coli* standard was having upon cheeses made in the United States and abroad, and it posed questions such as:

- What was the scientific basis for this change in regulation?
- Why was a more stringent *E. coli* standard warranted?
- What scientific evidence indicated that raw milk cheeses produced under current practices present a public health risk?
- Was it appropriate to apply the same microbiological standards to raw milk and pasteurized milk cheeses?
- To what extent did FDA consult with international organizations and producers who import cheese into the US?

This last question was particularly relevant since an intent of the FSMA was to harmonize US microbiological standards with those of our international trading partners.

In the response to those questions about *E. coli* standards, dated February 8, 2016, the FDA wrote:

> Regarding your questions about FDA's *E. coli* criterion for dairy products, non-toxigenic *E. coli* has long been used by the FDA and other public health regulatory agencies in the United States and other countries as an indicator organism for the presence of

> fecal contamination, which could indicate insanitary conditions in a processing plant.

The FDA offered four references in support of this and other statements in the letter. One of the references the FDA cited in preparing its scientific response for members of Congress was a 2004 article published in the journal *Food Microbiology* by Fook Ye Chye titled "Bacteriological Quality and Safety of Raw Milk in Malaysia." Was this really the best and most appropriate scientific evidence the FDA could cite for Congress in support of its *E. coli* standards that were imposing significant financial impacts on the domestic and imported artisan cheese industry?

That same day, on February 8, 2016, the FDA paused its *E. coli* testing. Here is the FDA's press release:

> Recently, cheesemakers have raised concerns suggesting that the FDA is applying safety criteria that may, in effect, limit the production of raw milk cheese without demonstrably benefiting public health. . . .
>
> Some question testing raw milk cheese for the presence of non-toxigenic *E. coli*, which has long been used by FDA and other public health agencies in the U.S. and other countries to indicate fecal contamination. Specifically, the concerns include the application of the test results and scientific foundation of these criteria.
>
> The FDA's reason for testing cheese samples for non-toxigenic *E. coli* is that bacteria above a certain level could indicate unsanitary conditions in a processing plant. Our surveillance sampling shows that the vast majority of domestic and imported raw milk cheeses are meeting the established criteria.
>
> Looking ahead, with the FSMA preventive controls rule now final, we will be taking another look at what role non-toxigenic *E. coli* should have in identifying and preventing insanitary conditions and food safety hazards for both domestic and foreign cheese producers.
>
> The FDA will also consider and update, as appropriate, the 2010 Compliance Policy Guide, which outlines safety criteria.

> Any changes will be informed by our engagement with stakeholders and experts on such issues as the use of a single bacterial criterion for both pasteurized and raw milk cheese, and the use of non-toxigenic *E. coli* as an indicator organism.
>
> The agency will continue to inspect cheese-making facilities and test for pathogens in domestic and imported cheese but, in the meantime, FDA is in the process of pausing its testing program for non-toxigenic *E. coli* in cheese. We will also continue working with all stakeholders to benefit from their expertise about safe cheese-making practices and achieve the mutual goal of food safety.

Somehow, in preparing its response to members of Congress, the FDA had neglected to mention the advice it had received from its own National Advisory Committee on Microbiological Criteria for Foods. In a November 17, 2014, draft document titled "Response to Questions Posed by the Department of Defense Regarding Microbiological Criteria as Indicators of Process Control or Insanitary Conditions," the Committee stated:

> The Committee believes that the best assessment of insanitary conditions is not necessarily finished product testing, but is typically best achieved through strategic evaluation of the production inputs, cleaning and sanitation practices and their efficacy, and the environmental monitoring and sanitation effectiveness monitoring data generated by the supplier facility as part of their preventive controls program. The verification test data from finished product testing, and even in-process product testing, when done, generally is simply too infrequent and too limited to make a reasonable prediction about the sanitary condition of a manufacturing establishment.[6]

So while the FDA was using non-toxigenic *E. coli* results in cheese as evidence of insanitary conditions in cheese, its own expert scientific advisory committee concluded that the finished product testing to which the FDA was subjecting cheeses was too infrequent or too limited to make a reasonable prediction about insanitary conditions.

On July 21, 2016, the FDA issued results of its raw milk cheese testing assignment. The FDA wrote:

> In 2014, the U.S. Food and Drug Administration (FDA) set out to collect and test cheese made from unpasteurized milk, also referred to as "raw milk cheese," aged 60 days as part of a new proactive and preventive approach to sampling with the ultimate goal of keeping contaminated food from reaching consumers.
>
> The FDA issued the raw milk cheese assignment in January 2014 along with two others (for sprouts and avocados) as the initial commodities under its new sampling model. As planned, the FDA collected 1,606 raw milk cheese samples (exceeding its target by 6 samples). The FDA designed its sampling plan such that if contamination of one percent or greater was present in the commodity, the agency would detect it. The agency closely monitored the assignment to gather lessons learned and make changes to the sampling if needed to address trends or food safety issues.
>
> Of the 1,606 raw milk cheese samples collected and tested, 473 samples (29 percent) were domestic samples, and 1,133 samples (71 percent) were of international origin. The FDA sought to design its sampling plan to approximate the ratio of domestically made versus imported product on the U.S. market but was unable to do so in this case because the federal government does not track production volume of raw milk cheese. Details on the assignment design are provided in the Sample Collection section of the summary report.
>
> The FDA tested samples for the presence of the pathogens *Salmonella*, *Listeria monocytogenes*, *E. coli* O157:H7 and Shiga toxin-producing *E. coli*, as well as for generic *E. coli*. The overall contamination rate for each of the pathogens was less than one percent, and the overall contamination rate for generic *E. coli* was 5.4 percent. While the prevalence for generic *E. coli* was comparatively high, it bears mention that it rarely causes illness even as it may signal insanitary processing conditions.
>
> Because the contamination frequencies among the pathogens were below one percent, the FDA was limited in its ability

> to detect differences in contamination rates based on the type of cheese or its origins (i.e., domestic vs. import), even with the large number of samples.
>
> In addressing the violative samples of domestic raw milk cheese, the FDA worked with the responsible firms to carry out recalls as appropriate and followed up with facility inspections. In addressing the violative samples of imported raw milk cheese, the FDA refused entries of raw milk cheese and placed the responsible firms/product on Import Alert 12-10. The FDA also worked with a regulatory partner in the European Union to address further follow-up of manufacturing locations abroad.
>
> *Listeria monocytogenes* in cheeses, particularly semi-soft varieties, remains a concern, as demonstrated by the nearly one percent contamination rate in semi-soft cheeses (see report Appendix B: Positive Findings by Bacterial Type). The FDA believes this contamination rate may be related to product handling practices or procedures. Given the serious public health implications of *Listeria monocytogenes* contamination associated with ready-to-eat foods, the FDA plans to continue to work with the cheese industry to identify and correct practices that lead to *Listeria monocytogenes* contamination in cheese.[7]

On September 15, 2016, Congressman Welch and 12 of his colleagues again sent a bicameral letter to the FDA's new Deputy Commissioner for Foods and Veterinary Medicine, Dr. Stephen Ostroff, expressing concern about FDA's *E. coli* standards. In response, on November 21, 2016, Dayle Cristinzio, Acting Associate Commissioner for Legislation, wrote:

> Fairly high percentages of both imported (93.9 percent) and domestic (96.2 percent) cheeses were found to not have levels of generic *E. coli* levels that would be of concern to the Agency. In other words, both imported and domestic cheeses had domestic generic *E. coli* levels that did not violate the criteria that were in place at the time of the sampling assignment, i.e., *E. coli* levels exceed 100 most probable number (MPN)/g in any single

> sample of cheese and/or did not exceed 10 MPN/g in three or more out of five samples.

I wrote to Senator Bernie Sanders's office on November 28, 2016, in response to the FDA's comments stating that the FDA has completely missed the point on this issue.

The reason so many cheeses were in compliance with new *E. coli* regulations is that cheesemakers tested products and released into commerce only those that met the new criteria. The criteria are without scientific merit and inconsistent with criteria that have been established internationally through *Codex Alimentarius* and ICMSF. The arbitrary FDA criteria are having a major financial impact on our domestic artisan producers as well as our international colleagues who produce raw milk cheese. Anecdotal information suggests that cheesemakers destroyed thousands of dollars' worth of cheese and spent thousands of dollars testing for meaningless standards. As a science-based agency, FDA has failed to show its scientific rationale for these standards and why US standards have not been harmonized with those of our international trading partners. The American Cheese Society is not a scientific organization. FDA has an obligation to justify to the scientific community how and why these criteria were developed, and that they are achieving food safety goals.

The FDA did not need to launch a food sampling pilot program; it had microbiological data on cheeses going back to 1998. While it was trying to explain to members of Congress that its *E. coli* standards were part of the FSMA, the CPG was changed in 2009, two years before the FSMA was signed into law. What else was going on to cause FDA to so drastically change *E. coli* standards in cheese?

When asked what power Congress has over the FDA, Welch said, "We can speak up and get attention for the cheesemakers." Welch also noted, "Congress has power over the FDA's budget."

CHAPTER 8

FDA's Assault on Artisan Cheese, Phase 2

Wooden Boards in Cheese Aging

Why, on February 8, 2016, did the FDA suspend *E. coli* testing of cheese? It was likely due to a combination of factors. One reason could have been that the FDA knew that its position on non-toxigenic *E. coli* could not be scientifically justified and was in conflict with guidance from the ICMSF (which makes you wonder why the FDA proposed this guidance in the first place). A second reason could have been due to fear over the way the FDA's proposed ban on wooden shelves in cheese aging was ultimately resolved.

By way of background, the use of wooden shelves in cheese aging is a centuries-old traditional practice. In addition to many AOP (Appellation d'Origine Protégée) and PDO (Protected Designation of Origin) cheeses that are required to be aged on wood, many US cheesemakers use wood in the aging of their artisan cheeses. In an effort to assist US artisan cheese producers, US scientific organizations supporting the artisan cheese industry began publishing information regarding the use of wood in cheese aging. A 2013 issue of *Dairy Pipeline*, published by the Center for Dairy Research (CDR) at the University of Wisconsin, featured an article, "Future Uses of Wooden Boards for Aging Cheeses," written by Bénédicte Coudé and Dr. Bill Wendorff. In the article, the authors stated:

Figure 8.1. Comté ages on wooden shelves. *Image courtesy CIGC*

> Most artisan cheesemakers feel that wooden shelves favor cheese rind development and improve the organoleptic qualities of aged cheeses thanks to the formation of a biofilm on the wood surface. Is this biofilm safe? Might it favor the development of spoilage and potential pathogenic bacteria? The purpose of this review is to look at the benefits of wooden boards as well as the potential concerns and how to avoid them.[1]

In January 2014, the American Society for Microbiology, the world's largest organization of biological scientists, had published in *Microbiology Spectrum* a chapter from a then upcoming book titled *Cheese and Microbes*. The chapter, "Wooden Tools: Reservoirs of Microbial Biodiversity in Traditional Cheesemaking," was authored by three eminent European scientists: Drs. Sylvie Lortal and Florence Valence of the INRA (National Institute for Agricultural Research), and Giuseppe Licitra of CoRFiLaC (Consorzio per la Ricerca nel Settore della Filiera Lattiero Casearia, or

Consortium for Research in the Dairy Sector).[2] The authors indicated that wooden shelves are used for the ripening of approximately 500,000 tons of cheese in Europe annually; with approximately 350,000 tons of that cheese produced in France alone. Many well-known PDO cheeses, including Comté, Reblochon, Beaufort, Muenster, Cantal, and Roquefort, require aging on wood. The use of wooden tools is mandatory for some PDO cheeses, for example, the "gerle" (wooden vat) in the production of Salers cheese, and the tina wood vat in Ragusano cheese production. Included in the EU directive 96/536/EC is reference to the use of spontaneous microflora and specific cheesemaking technology referring to processes that avoid the use of starters and reflect traditional methods in the areas of PDO cheese production. Many cheesemakers consider the use of wooden tools essential for improving the organoleptic and typical characteristics of PDO cheeses.

It was therefore surprising to cheesemakers when, in 2014, the FDA issued guidance on the use of wooden shelves in cheese aging. On January 14, 2014, Rob Ralyea, a senior extension associate in the Department of Food Science at Cornell University, sent a message to the artisan cheesemaking community indicating that he had put an inquiry into the regional FDA, who passed it along to the national FDA, on the agency position regarding the use of wooden boards in cheese caves. Ralyea indicated that this issue had arisen in several New York locations, not from the New York State Department of Agriculture and Markets, Division of Milk Control and Dairy Services personnel, but from FDA inspectors going to cheese plants under the direction of the Food Safety Modernization Act (FSMA). Ralyea indicated to the cheesemaking community that while he did not have an official response in hand, a highly reliable source had informed him that a final draft was being done and the FDA would not allow any wood in cheese aging rooms as a product contact surface.

The reference to "FDA inspectors going to cheese plants" comes from an inspection of an artisan cheese facility located in upstate New York that was conducted by the FDA on July 9–12, 2012. As a result of the inspection, the FDA found *Listeria monocytogenes* on one wooden board used to age cheese and issued a warning letter to Finger Lakes Farmstead Cheese Company.[3] In 2013, *L. monocytogenes* was found again, and on January 22, 2014, Finger Lakes Farmstead was issued an injunction and entered into a consent decree.[4] The company was forced to cease operations, destroy its inventory, hire an independent laboratory and sanitation expert, train employees, and develop a *Listeria* control plan.

On January 22, 2014, Ralyea informed cheesemakers that the FDA had released its policy on wooden shelves for cheese aging. Here is the official statement, in part, from the FDA Center for Food Safety and Applied Nutrition (CFSAN):

> Microbial pathogens can be controlled if food facilities engage in good manufacturing practice. Proper cleaning and sanitation of equipment and facilities are absolutely necessary to ensure that pathogens do not find niches to reside and proliferate. Adequate cleaning and sanitation procedures are particularly important in facilities where persistent strains of pathogenic microorganisms like *Listeria monocytogenes* could be found. The use of wooden shelves, rough or otherwise, for cheese ripening does not conform to cGMP requirements, which require that "all plant equipment and utensils shall be so designed and of such material and workmanship as to be adequately cleanable, and shall be properly maintained." 21 CFR 110.40(a). Wooden shelves or boards cannot be adequately cleaned and sanitized. The porous structure of wood enables it to absorb and retain bacteria, therefore bacteria generally colonize not only the surface but also the inside layers of wood. The shelves or boards used for aging make direct contact with finished products; hence they could be a potential source of pathogenic microorganisms in the finished products. . . .
>
> The primary concern for cheese manufacturers should be prevention of cheese contamination with pathogens. Pathogenic microorganisms are not inherent natural contaminants of cheeses, therefore the sanitation of a cheese processing plant's equipment and environment plays an important role in preventing pathogen contamination.

It appears that the FDA's official position on wooden shelves was written in direct response to the *Dairy Pipeline* article by Coudé and Wendorff.

According to the document properties, it was created by Dr. Obianuju Nsofor, a microbiologist at CFSAN's Dairy and Egg Branch.

Of course, cheesemakers aging cheese on wood and those companies like Whole Foods and Wegmans selling cheeses aged on wood were taken aback. Cheesemakers are required to have a robust microbiological monitoring system, and they must track shelving with lot codes so as to comply with traceability requirements for recalls in compliance with programs such as Safe Quality Food (SQF), a globally recognized food safety and quality certification program that is a requirement of many retailers.

On June 4, 2014, Ralyea again advised the artisan cheese community that the FDA had inspected some artisan cheesemakers in New York State in 2014 and cited the inspected establishments for the practice of using wooden boards in cheese aging. The New York State Department of Agriculture and Markets, Division of Milk Control and Dairy Services has regulatory authority for dairy product safety, and it had allowed the use of wooden boards for aging of cheese as long as the boards were maintained in a clean and sanitary condition, consistent with Code of Federal Regulations (CFR) 110 requirements.

Ralyea also explained to the artisan cheese community that according to the FDA, for the FSMA compliance, cheesemakers importing cheese to the United States would be subject to the same rules and inspections as US cheesemakers. He stated, "Therefore, it stands to reason that if an importer is using wood boards, the FDA would keep these cheeses from reaching our borders until the cheesemaker is in compliance." The FDA states that it will be consistent in its application to both domestic and foreign producers, even though Health Canada has stated that using wood boards for aging is not an issue, nor is it being changed, and the EU authorizes and allows the use of wood in cheese aging.

The New York State Department of Agriculture and Markets had also sought clarification from the FDA on the use of wooden boards for cheese aging. It received a response from Monica Metz, the branch chief for the FDA Center for Food Safety and Applied Nutrition's Dairy and Egg Branch, who wrote: "The use of wooden shelves, rough or otherwise, for cheese ripening, does not conform to Current Good Manufacturing Practices (cGMP) requirements, which require that 'all plant equipment and utensils shall be so designed and of such material and workmanship as to be adequately cleanable, and shall be properly maintained.'" The FDA quotes 21 CFR 110.40(a) as its reference. The FDA also noted that this was not a change in policy, merely proper enforcement of a policy that was already in place.

Vermont cheesemakers, worried about the status of their cheese that was being aged on wood, contacted the office of Congressman Peter Welch to obtain clarification on the wooden boards issue. On March 27, 2014, Phil Broadbent of the FDA's Office of Legislation, sent Patricia Coates, then aide to Congressman Welch, two articles that supported FDA's position that wooden boards cannot be adequately cleaned and sanitized:

> I inquired with FDA's food center [the Center for Food Safety and Applied Nutrition] regarding your question about the use of wooden shelves for the storing or aging of cheeses, and I was told they are not permitted and never have been. The use of wooden shelves for cheese ripening does not conform to current Good Manufacturing Practice (GMP) regulations, which require that "all plant equipment and utensils shall be so designed and of such material and workmanship as to be adequately cleanable, and shall be properly maintained." See 21 CFR 110.40(a).
>
> Wooden shelves or boards cannot be adequately cleaned and sanitized. The porous structure of wood enables it to absorb and retain bacteria, therefore bacteria generally colonize not only the surface but also the inside layers of wood. This has been borne out by recent academic studies showing that the bacterium *Listeria monocytogenes* survived cleaning and sanitation on wooden shelves used for cheese ripening.

A review of the two articles sent to Congressman Welch by the FDA reveals that the FDA's interpretation of the articles was incorrect. The FDA states: "Recent publications by Zangerl et al. 2010 show that *L. monocytogenes* survived cleaning and sanitation on wooden shelves for cheese ripening." However, an actual quote from the Zangerl article stated: "The present study shows that the use of wooden shelves does not affect the hygienic safety of cheeses if such shelves are in good repair and are thoroughly cleaned and sanitized by heat treatment. Therefore, there is no reason to replace wood employed in cheese ripening process with other materials."

As for the second article, the FDA concludes, "Thus the paper (Mariani) does not support the proposition for which it was offered, viz. that wooden shelves prevent contamination of all cheeses with *L. monocytogenes*." The actual article by Mariani states: "All together, our results suggest that the biocontrol of pathogens multiplication on wooden shelves by resident biofilms should be considered for the microbiological safety of traditional ripened cheeses. . . . Whatever the conditions tested, we observed a significant reduction in the population of the pathogen on native wooden samples after 12 days of incubation."[5]

It is terrifying to think that, with the stroke of a pen, the FDA could wipe out centuries of traditional practice. It is even more frightening to think that this could be accomplished through the incorrect interpretation of scientific studies. It is notable that the FDA's guidance omitted a recommendation by Zangerl of heating cleaned boards to 176°F (80°C) for 5 minutes or 149°F (65°C) for 15 minutes to eliminate pathogens.[6] This omission challenges FDA's notion that wooden boards cannot be adequately cleaned or sanitized. Furthermore, the FDA's guidance ignores ongoing efforts of cheesemakers to monitor wood shelving for presence of *Listeria* consistent with its own FSMA Proposed Rule for Preventive Controls in Human Food.

The primary concern for cheesemakers with the FDA's guidance was that elimination of wooden shelves in cheese aging poses significant adverse financial consequences, as well as broader implications for the international trade of AOP and PDO cheeses. It is curious that this issue appeared as the United States was preparing for negotiations over the Transatlantic Trade and Investment Partnership (T-TIP) agreement. A sticking point in the negotiations of this agreement arose over geographical indications (GI) that protect the names of European cheese varieties. Some US cheesemakers, represented by the Consortium for Common Food Names (CCFN), objected to these GI protections. By banning wood in cheesemaking, the FDA could effectively ban the importation of European cheeses such as Parmigiano Reggiano, Comté, Reblochon, Beaufort, Muenster, Cantal, and Roquefort, among others.

Other governments have acted in more transparent ways. As previously discussed, Food Standards Australia New Zealand (FSANZ) established regulations regarding the sale of Roquefort cheese in Australia. Anyone who has visited Australia knows that its food safety standards are some of the most stringent in the world. At the time, Australia did not allow the manufacture of raw milk cheese in Australia. The French government filed an application to permit the sale in Australia of Roquefort cheese, a raw milk cheese that is aged on wooden boards in limestone caves located in the high plateaus of France's Causse du Larzac in the town of Roquefort-sur-Soulzon. In response to this request, FSANZ undertook a comprehensive risk assessment on this issue, consistent with the *Codex Alimentarius* principles, which guide safety assessments of foods in global commerce, and published the results of this assessment in 2005.[7] FSANZ acknowledged in the risk assessment that while normal food manufacture premise conditions are not applied, the controls and checks used during Roquefort manufacture are adequate to ensure product safety. The risk assessment goes on to state that the cave is an essential part of the process for the manufacture of Roquefort: "While approximately 14 days of cheese maturation occurs in caves on wooden boards where normal food manufacture premise conditions are not applied, the controls and checks of the cheese entering the cave maturation and the controls and checks following this period provides sufficient confidence that Australian requirements can be met." As noted in results from an audit conducted at the caves, "The condition of the caves did not comply with the standards which would be expected in Australian storage facilities, however, these caves form an essential part of the process for the development of Roquefort cheese. Monitoring of the cave environment and product subsequent to cave storage ensures that there is minimal risk of contamination of product, and identification of product should it become contaminated."

As Jeanne Carpenter wrote on June 7, 2014, on her blog, *Cheese Underground*:

> Wisconsin cheesemaker Chris Roelli says the FDA's "clarified" stance on using wooden boards is a "potentially devastating development" for American cheesemakers. He and his family have spent the past eight years re-building Roelli Cheese into

a next-generation American artisanal cheese factory. Just last year, he built what most would consider to be a state-of-the-art aging facility into the hillside behind his cheese plant. And Roelli, like hundreds of American artisanal cheesemakers, has developed his cheese recipes specifically to be aged on wooden boards.

"The very pillar that we built our niche business on is the ability to age our cheese on wood planks, an art that has been practiced in Europe for thousands of years," Roelli says. Not allowing American cheesemakers to use this practice puts them "at a global disadvantage because the flavor produced by aging on wood can not be duplicated. This is a major game changer for the dairy industry in Wisconsin, and many other states."

As if this weren't all bad enough, the FDA has also "clarified"—I'm really beginning to dislike that word—that in accordance with FSMA, a cheesemaker importing cheese to the United States is subject to the same rules and inspection procedures as American cheesemakers.[8]

Following social media posts by angry cheesemakers, warning consumers of the potential impacts of FDA's wooden board ban, the FDA issued a Constituent Update on June 11, 2014:

Recently, you may have heard some concerns suggesting the FDA has taken steps to end the long-standing practice in the cheesemaking industry of using wooden boards to age cheese. To be clear, we have not and are not prohibiting or banning the long-standing practice of using wood shelving in artisanal cheese. Nor does the FDA Food Safety Modernization Act (FSMA) require any such action. Reports to the contrary are not accurate.

The agency's regulations do not specifically address the use of shelving made of wood in cheesemaking, nor is there any FSMA requirement in effect that addresses this issue. Moreover, the FDA has not taken any enforcement action based solely on the use of wooden shelves.

> At issue is a January 2014 communication from the agency's Center for Food Safety and Applied Nutrition to the New York State Department of Agriculture and Markets' Division of Milk Control and Dairy Services, which was sent in response to questions from New York State.
>
> The FDA recognizes that this communication has prompted concerns in the artisanal cheesemaking community. The communication was not intended as an official policy statement, but was provided as background information on the use of wooden shelving for aging cheeses and as an analysis of related scientific publications. Further, we recognize that the language used in this communication may have appeared more definitive than it should have, in light of the agency's actual practices on this issue.
>
> The FDA has taken enforcement action in some situations where we have found the presence of *Listeria monocytogenes* at facilities that used such shelving. Since 2010, FDA inspections have found *Listeria monocytogenes* in more than 20 percent of inspections of artisanal cheesemakers. However, the FDA does not have data that directly associates these instances of contamination with the use of wood shelving.
>
> In the interest of public health, the FDA's current regulations state that utensils and other surfaces that contact food must be "adequately cleanable" and "properly maintained." Historically, the FDA has expressed concern about whether wood meets this requirement and these concerns have been noted in its inspectional findings. However, the FDA will engage with the artisanal cheesemaking community, state officials and others to learn more about current practices and discuss the safety of aging certain types of cheeses on wooden shelving, as well as to invite stakeholders to share any data or evidence they have gathered related to safety and the use of wood surfaces. We welcome this open dialogue.

In response to this Constituent Update, staff in the office of Congressman Peter Welch composed the memo "Cheese Lovers of the House Unite!: Stop the FDA from Banning Centuries-old Cheese Making Practice," which they sent to their House colleagues. The memo stated the following:

> Since the time of Adam and Eve, cheese makers have aged cheese on wood boards and wood shelves. Wood allows the cheese to breathe and develop its tangy and rich flavor during the aging process. In Europe, cheese makers are required to use wood shelves to age the product to develop the proper texture and flavor.
>
> Astonishingly, the Food and Drug Administration has begun a "crackdown" on America's small artisan cheesemakers for using wood shelves in the aging process. Without evidence to support its enforcement, the agency has cited the risk of contamination as the cause for its overreach. . . .
>
> This bureaucratic overreach by the FDA is a solution in search of a problem. Artisan cheesemakers already have rigorous protocols in place to assure the safety of their product. Instead of banning a centuries-old aging process and triggering a possible trade war with Europe, the FDA should take a deep breath and work collaboratively with food scientists and cheese makers to ensure their products meet the high standards expected by cheese loving consumers around the world.
>
> The burgeoning artisan cheese industry is made up of small business entrepreneurs dedicated to producing world-class American cheeses and creating the good jobs that go with it. Please support our amendment to stop the FDA in its tracks from doing serious harm to these small businesses.

Congressman Welch, Democrat from Vermont, also reacted to the FDA clarification on June 12, 2014, noting on Twitter: "The FDA's right hand doesn't know what the left hand is doing. Which FDA should cheese makers listen to? We will not back down."

There is much discussion about the inability of members of Congress to work collaboratively to solve problems important to the American people. In an amazing display of bipartisanship, the Welch amendment to H.R. 4800, the Agriculture, Rural Development, Food and Drug Administration and Related Agencies Appropriation Act of fiscal year 2015, inserts at the end of this bill the following:

"Sec. ll. None of the funds made available by this Act may be used to establish, implement or enforce any prohibition against aging or ripening cheese on wood under section 110.40 of Title 21, Code of Federal Regulations."

The amendment was jointly sponsored by Peter Welch and Paul Ryan, Republican from Wisconsin. Congressman Welch held a press conference on June 18, 2014, with members of Vermont's cheese industry to provide an update on this issue and its resolution.

Gregory McNeal wrote about the wooden boards issue in a 2014 article for *Forbes* titled "FDA Backs Down in Fight over Aged Cheese":

> This is also a lesson for people in other regulated industries. When government officials make pronouncements that don't seem grounded in law or policy, and threaten your livelihood with an enforcement action, you must organize and fight back. While specialized industries may think that nobody cares, the fight over aged cheese proves that people's voices can be heard.[9]

The public debate over wooden boards in cheese aging, and the FDA's response, was driven by activism. Perhaps the FDA did not want its budget threatened in a similar kerfuffle over its proposed *E. coli* standards.

Sadly, the FDA's attempt to ban the use of wooden boards in cheese aging came as no surprise to the artisan cheese community. It had been warned that the FDA was about to make this change in 2012. In August 2012, the American Cheese Society held its annual conference in Raleigh, North Carolina. In January 2012, Bill Graves of Dairy Management Inc. (DMI) had obtained an agreement from John Sheehan, director of FDA's Division of Plant and Dairy Food Safety, to be part of a panel at ACS that he had organized titled "Working Proactively to Assure Cheese Safety." The description of the session, as submitted to ACS, was as follows:

> Join this dynamic panel to get an in-depth look at how decision making at one end impacts the entire cheese chain from cheesemaker to distributor to retailer to consumer. John Sheehan from

> the FDA will provide an overview of current risk assessments, changing requirements, and results of surveillance activity including the impact of these on regulations . . .

Unfortunately, Sheehan reached out to the American Cheese Society (ACS) and informed it that, due to a scheduling conflict, he would be unable to attend the ACS meetings. It was therefore a great surprise when on July 31, he appeared in Raleigh to meet with members of the Innovation Center for U.S. Dairy's artisan/farmstead food safety team. Sheehan told the small assembled group, of which I was a part, that US producers would not be allowed manufacturing practices such as those in Europe ("we're getting rid of all the wood"), further stating that imported producers would be inspected and if violations occurred (per FDA), product would not be allowed into the country.

It also came as no surprise that the day following Congressman Welch's wooden boards press conference, FDA inspectors began collecting cheese samples from a company that had participated in the press conference.

CHAPTER 9

Listeria and the Soft Cheese Risk Assessment

In 2009, rather than its focus on non-toxigenic *E. coli* standards, the FDA should have been focused on control of *Listeria* in all US dairy plants, both industrial and artisan. By way of background, *Listeria monocytogenes* is an organism that is widely distributed in nature. It has been isolated from diverse environmental sources including soil, sewage, water, green plant material, fresh produce, and domesticated animals including ruminants, which is why this pathogen is of significance on dairy farms and in dairy environments. The association between poor-quality silage feeding and development of listeriosis in dairy cows, sheep, and goats is well recognized. Infected animals display neurological impairments and may circle endlessly in one direction, causing listeriosis to be referred to by veterinarians as circling disease. *L. monocytogenes* possesses characteristics that make it very tolerant of a wide range of environmental conditions, and this complicates control of this pathogen.[1] Most important, *L. monocytogenes* is one of very few human pathogens that has adapted to growth at refrigeration temperatures, which is why it poses challenges to the safety of cheeses. During refrigerated storage in some cheeses, low initial levels of *L. monocytogenes* can grow and multiply to high levels that can cause human illness. This pathogen can also display resistance to salt, acid, sanitizers, and heat. *L. monocytogenes* grows best at neutral pH ranges and in high-moisture environments, which is why fresh soft and soft-ripened cheeses support growth of this pathogen to high levels that can pose a risk to public health, especially for products that are stored under refrigeration for extended time periods. In contrast, in hard cheeses such as Cheddar, Swiss (Emmental, Gruyère), and Italian Grana cheeses, factors such as low pH and low moisture interact to create a hostile environment for this pathogen, preventing growth and accelerating declines in populations levels during aging.

A number of notable cheese-associated outbreaks of listeriosis have occurred around the world. We have already discussed the 1985 Mexican-style soft cheese outbreak and the Vacherin Mont d'Or outbreak between the years 1983 and 1987. An outbreak linked to Brie de Meaux, a soft-ripened raw milk cheese, occurred in France in 1995.[2] Twenty people developed listeriosis and four fetal deaths were reported. In 2000, illegally produced Mexican-style soft cheese (*queso fresco*) resulted in 13 cases of illness, 11 perinatal infections, and 5 stillbirths in North Carolina.[3] During 2009 and 2010, a large listeriosis outbreak involving 34 cases of illness and 8 deaths was reported in Austria, Germany, and the Czech Republic.[4] This outbreak was linked to Quargel, a traditional Austrian acid curd, red smear cheese made from pasteurized milk. An outbreak linked to imported Ricotta Salata cheese occurred in the United States in 2012, resulting in 22 cases of illness and 4 deaths. *L. monocytogenes* is of primary concern to regulatory officials, as it is a leading cause of death due to foodborne illness, and infections result in high hospitalization rates. It is a serious human pathogen and its presence in foods must be controlled.

Vermont is a wonderful state that enjoys a small population and a vibrant rural culture where dairy production dominates agriculture. Community is strong in Vermont, and neighbors are expected to help neighbors. When farmers or cheesemakers had questions about dairy product safety, it was not unusual for them to stop by my office or call to receive information. We built up a lot of trust with this wonderful dairy community, who were willing participants in research that allowed us to learn more about the ecology of *Listeria* and ways to control this dangerous pathogen. I cannot say enough positive things about the Vermont Agency of Agriculture, Food and Markets, whose leaders provided excellent support and guidance to the Vermont dairy industry, including the Vermont artisan cheese sector.

In 2008, with funding from the USDA Rural Business Enterprise Grants program, my research group conducted a comprehensive risk assessment for 16 Vermont farmstead cheesemakers.[5] We visited farms that produced raw milk specifically for use in artisan cheese production. We collected milk and analyzed it for bacterial pathogens. We also conducted an environmental analysis to determine if *Listeria* was present in the dairy processing environment where cheese was being made. We found that *Staphylococcus aureus* was the most common pathogen isolated from raw milk. We also isolated *L. monocytogenes* from environmental sites that included floors, drains, milk cans, and equipment. Our team shared the microbiological results with cheesemakers and suggested corrective actions (alteration of traffic flow patterns, operating procedures to eliminate cross-contamination, and improved sanitation) to eliminate the presence of *Listeria*. We came back a few months later after corrective actions had been made to determine if our guidance had been effective. We again performed environmental sampling to see if we could find *Listeria*. In all cases, corrective actions were effective, as *Listeria* had been eliminated from the previously contaminated sites. The goal of our research had not been simply focused on finding *Listeria*; the goal had been to identify sites of contamination and then develop strategies to control or eliminate this pathogen in artisan cheese production environments.

Again in 2011–2012, my group developed risk reduction protocols by conducting on-site visits to each of 10 participating Vermont cheesemakers. We spent 2 days with each cheesemaker. On the first day, a comprehensive review of the cheesemaking process, from milking to aging, was conducted to allow a flow diagram of the cheesemaking process to be developed. Using a Hazard Analysis and Critical Control Points (HACCP) approach, we identified critical control points in the process. The type of cheese being manufactured, an assessment of risk (high risk to low risk, dependent upon cheese characteristics), and physical notes about the processing facility (condition, layout, traffic flow) were compiled. Next, the cheesemaker conducted cheesemaking with participation from the technical team from the Vermont Institute for Artisan Cheese (VIAC). We recommended a system for achieving process control by identifying the key parameters that needed to be routinely measured during cheesemaking (such as pH, titratable acidity, salt-in-moisture, percent moisture) based upon the federal standard of identity for the cheese being made.

We collected samples of milk, curds, and whey during cheese manufacture, along with the finished cheese, and analyzed samples for standard plate count (SPC), coliforms, and somatic cell count (SCC), along with pathogens consisting of *L. monocytogenes*, *Salmonella*, *E. coli* O157:H7, and *S. aureus*. We also collected environmental swab and sponge samples from targeted areas in the cheese manufacturing facility (floor drains, floors, vats, tables, carts, squeegees/floor mops) and analyzed these for presence of *L. monocytogenes*.

We shared the results of the microbiological analysis with the cheesemaker and made recommendations to improve the cheesemaking process, such as changes in the make process; physical layout of the facility and reorientation of foot traffic; changes in sanitation; the need for protective clothing—gowns, hairnets, gloves—and proper hand washing and sanitization; implementation of hygienic zoning; improvements in milk quality, and so forth.

Once the cheesemaker implemented the recommendations, we returned for a second visit, at which time cheesemaking was again conducted with the VIAC technical team and a comprehensive microbiological analysis was conducted. We compared our results between visit 1 and visit 2 to determine if the recommendations made by the VIAC technical team resulted in risk reduction, improved process control, and improved cheese safety and quality. In every instance, our recommendations resulted in elimination of *Listeria*, as it was never found in our subsequent sampling. This work proved to be timely; in April 2010, FDA's Center for Food Safety and Applied Nutrition (CFSAN) issued its Request for Inspections and Environmental Sampling for *Listeria* at Soft Cheese Firms. The assignment (FY 2010 Soft Cheese Assignment DFPG #10-04) instructed FDA districts around the country to collect environmental samples from areas in facilities that were not generally sampled.[6] The purpose of these inspections was to determine whether or not *Listeria* was present in the food processing environment in firms producing soft cheese. What this meant for artisan cheesemakers was teams of three FDA inspectors showing up unannounced and subsequently spending between 1 and 3 days at each facility collecting a total of 100 to 300 swabs from all areas in the cheesemaking facility. Between 2010 and 2011, the FDA conducted environmental surveillance of US cheesemakers in 27 states producing soft cheese (154 plants total, 41 artisan producers), and found that 31 percent of the plants that were tested had positive environmental findings for *L. monocytogenes*. Of artisanal firms, 20 percent had positive environmental findings compared to 27 percent of nonartisanal firms. I am very pleased to state that no Vermont cheesemaker visited by FDA had positive environmental samples for *Listeria*. Educational and mitigation efforts are critical and can provide interventions for control of this extremely dangerous human pathogen.

Other cheesemakers, such as Finger Lakes Farmstead, were not so fortunate, and the finding of *L. monocytogenes* led to recalls in some instances and outbreak investigations in others. Clearly, educational efforts were having an impact on *Listeria* control by artisan cheesemakers. Since we had found a successful formula, we were more than happy to share the results of our work with broader audiences. The Innovation Center for U.S. Dairy, according to its website, "is a forum that leverages collective power in the dairy industry to address the changing needs and expectations of consumers through a framework of shared best practices and accountability."[7] Food safety in general is a topic of interest for the Innovation Center, and the safety of artisan and farmstead cheese was one of the issues targeted for focus by this group. The Innovation Center, through its partner Dairy Management Inc., had invited representatives of the academic dairy centers across the country to a meeting in Chicago to discuss strategies for improving artisan and farmstead cheese safety. At one of the meetings, I suggested that, rather than reinventing new strategies, why not use the information that VIAC had already developed to share with other groups?

My research team worked with the Innovation Center for U.S. Dairy to share our *Listeria* control information. A number of highly successful Dairy Innovation Center / VIAC Artisan Safety Workshops were held at locations around the country. The workshops were well attended by artisan and farmstead cheesemakers and covered everything from basic information about the pathogens that cheesemakers needed to control to information about sanitation and facility design to strategies for implementing a food safety plan. Some of the workshops were held in conjunction with the annual meetings of the American Cheese Society, which made cheesemaker attendance more feasible.

All of the workshops were extremely well received. What made our workshops different from other dairy safety workshops offered through extension programs at universities across the country was our focus on the specific needs of small-scale food producers. It is easy to convey scientific information to other scientists; we speak the same language. It is much more of a challenge to communicate scientific concepts to people who may not have extensive scientific backgrounds. We found a way to translate our scientific language in a manner that was understandable and could be readily implemented by cheesemakers. Through our work at VIAC, many of the "cheesemakers" who enrolled in our educational classes were there to learn about cheesemaking as a second career, having finished highly successful first careers in banking, law, engineering, journalism, information technology, or medicine. These were highly educated, skilled, and savvy individuals who wanted a "deep dive" into cheesemaking. The information they needed had to be efficiently presented and comprehensive, but not "dumbed down." We were successful in delivering information right on target to these groups.

Feedback we received through the Innovation Center was that the FDA and state dairy inspectors were not always knowledgeable about artisan cheesemaking, making inspections more difficult for everyone involved. We were asked how we felt about having cheesemakers and inspectors in the same room to attend the same trainings. As a 35-year veteran university professor, I believe strongly that education should be available to all, and we were more than agreeable to having dairy inspectors attend our courses. Late in October 2013, a workshop titled "Food Safety and Hygiene in Artisan/Farmstead Cheesemaking," cosponsored by the Georgia Department of Agriculture, the Innovation Center for U.S. Dairy, and Whole Foods, was held in Georgia.

To our disappointment and dismay, the day after the workshop, every cheesemaker in attendance, along with two Whole Foods Market stores closest to the base location of the inspectors (Georgia and Northern Florida), received an inspection, along with detention of their cheeses. The notion that attendance at, or sponsorship of, an educational workshop could subject an artisan cheesemaker or cheesemonger to regulatory scrutiny defeated the very purpose for which these workshops were intended. It was never clear how regulators obtained information on the attendees at this workshop. However, as I indicated to my colleagues at the Innovation Center for U.S. Dairy when I attended my last meeting with this organization, trust had now emerged as an issue, and continuation of these important safety workshops could no longer be justified.

CHAPTER 10

Geographical Indications

What is the impact of FDA regulatory challenges on the future growth of the US artisan cheese industry? On AOP (Appellation d'Origine Protégée) and PDO (Protected Designation of Origin) cheeses? A curious and uncomfortable relationship exists between the FDA and the mainstream US dairy industry, and evidence points to a collaboration working against the interests of the US artisan cheese industry. During the time when the FDA was waging its punishing war against US artisan cheesemakers, was it really helping US dairy industrial giants fight a trade war with the EU over protections for its traditional cheeses? Were the FDA's efforts to initiate the Domestic and Imported Cheese and Cheese Products Compliance Program in 1998 and establish stringent *E. coli* criteria in 2009 and 2010 and ban the use of wood in cheese aging in 2014 all steps designed to control safety threats, or were they intended to be used as trade barriers against EU-produced cheeses? Were US artisan cheesemakers erroneously caught in the crossfire, or were these efforts intended to target and close down US artisan cheese production as well? The timeline of the FDA's punishing attempts at regulatory activity toward the artisan cheese community tracks closely with the implementation of protections that the EU put in place to safeguard its traditional foods, and the disputes that have followed at the World Trade Organization (WTO).

On July 14, 1992, the EU Council Regulation (EEC) No. 2018/92 on the protection of geographical indications and designations of origin established two types of geographical indication (GI) designations for agricultural products and foodstuffs.[1] The first, Protected Designation of

Figure 10.1. The 60th anniversary of the Comté PDO. *Image courtesy CIGC*

Origin (PDO), indicates that the quality of a product is essentially *due to that region*. An example of a PDO cheese is Comté, which is produced in the Franche-Comté region that borders Switzerland. The inherent characteristics of Comté are derived from the air, climate, land, native plant species, and technological expertise of regional producers. The second, Protected Geographic Indication (PGI), indicates that the quality of a product is *attributable to that region*. French Gruyère is an example of a cheese, made in a defined production zone, that obtained a PGI designation in 2012. Geographical indications (GIs) enjoy protection under the TRIPS Agreement, a multilateral agreement on intellectual property protection, and fall under the jurisdiction of the World Trade Organization. Over 4,000 food products globally benefit from geographical indications, designations designed to preserve the methods, skills, and regional specificity required of these products. Italy holds the majority of geographical indications, with 269 protected food products and 523 wines. France holds more than 600 GI protections, the majority in place for wines such as Bordeaux and Champagne. Products including Parma ham, Colombian coffee, Mexican tequila, and Scotch whisky are just a few examples of other products that enjoy GI protections. A long list of traditional cheeses from Europe carry GI designations, including Comté, Brie, Camembert, and Roquefort.

Countries outside the EU have complained that the PDO status of these products prevents lawful competition in the marketplace. As Dick Groves from the publication *Cheese Reporter* writes: "US-EU dairy trade is currently pretty much a one-way street, with EU dairy exports to the US running close to $1.5 billion over the last couple of years, while US dairy exports to the EU reached a recent high of $143 million back in 2013."[2] Currently, traditional cheeses made in the EU account for a majority of the dairy trade imbalance: In 2016, the EU shipped $972 million of cheese to the United States, while the United States shipped just $6 million of cheese to the EU.[3]

In a dispute filed with the WTO on June 1, 1999, the United States requested consultations with the European Commission over disputes of alleged lack of protection of trademarks and geographical indications for agricultural products and foodstuffs in the EC. The United States contended that Council Regulation (EEC) No. 2081/92, as amended, failed to provide national treatment for geographical indications, and did not provide adequate protection to preexisting trademarks that were similar or identical to a geographical indication.[4] US producers expressed concern that the EU position of protecting its traditional products would force US producers of these products to rename them, thereby creating confusion for US consumers, while forcing American companies to spend extensive resources in the rebranding of these products. In response, the United States led a campaign to oppose the EU's use of GI restrictions as protectionist strategies that create trade barriers.

Many raw milk cheeses imported into the United States enjoy PDO status. By requiring mandatory pasteurization of all milk used for cheesemaking, or banning the use of wood in cheese aging, the United States can effectively eliminate importation of PDO cheeses because the PDO requires them to be manufactured from raw milk and/or aged on wood. Was the FDA helping US industrial dairy producers retaliate against the EU by imposing these regulations? During the debacle over use of wooden boards in cheese aging, I posed this question to our Vermont congressional staff, who furnished a March 11, 2014, letter written to USDA Secretary Tom Vilsack and US Trade Representative Michael Froman. The letter, authored by Senators Chuck Schumer and Pat Toomey and signed by 53 members of the US Senate, commended Secretary Vilsack and Michael Froman for their work to fight growing GI restrictions promoted by the EU, which they viewed as a "trade barrier of great concern to dairy and other food manufacturers." They expressed concern that in the Transatlantic Trade and Investment Partnership (T-TIP) negotiations, the EU was seeking to impose restrictions on common food names. As stated in the letter, "In the states that we represent, many small or medium-sized family owned farms or firms could have their business unfairly restricted by the EU's push to use geographical indications as a barrier to dairy trade and competition." The T-TIP negotiations are currently stalled, but GIs have once again become an issue in negotiations of other trade agreements.

This was not a new issue is 2014. In fact, on July 22, 2003, the House Agriculture Committee of the US Congress held hearings to examine the possible impacts of the implementation of geographic indication protections proposed in World Trade Organization negotiations. The most contentious issue associated with the GI protections was Europe's attempt to protect not only the traditional name of a product—such as *Parmigiano Reggiano*—but also the common name *Parmesan*. Michael Pellegrino, Vice President of Marketing and Strategy of the Kraft Cheese Division of Kraft Foods North America, gave testimony before this committee. As Mr. Pellegrino testified:

> Kraft is one of the world's largest producers of Parmesan cheese. Our ability to continue to sell Parmesan cheese and other Kraft products is jeopardized by initiatives being aggressively advanced by the European Union in the ongoing Doha Round [of trade talks]. . . . This year, Kraft will manufacture and market about 60 million pounds of Parmesan cheese under the Kraft and DiGiorno brands, absorbing nearly 1 billion pounds of US-produced milk. We regard the threat to our businesses, and to those of other US and non-EU food processors and producers, as real, substantial, and immediate. Denied use of these names, Kraft would have to convince each of our consumers that the Kraft grated cheese being sold under an unfamiliar name is the same high-quality cheese they have been serving for decades as Kraft Parmesan cheese.

A look at the US Code of Federal Regulations Title 21, Part 133.165 reveals the standard of identity for "parmesan and reggiano" cheese. The section begins: "(a) Parmesan cheese, reggiano cheese, is the food prepared from milk and other ingredients specified in this section, by the procedure set forth in paragraph (b) of this section, or by another procedure which produces a finished cheese having the same physical and chemical properties as the cheese produced when the procedure set forth in paragraph (b) of this section is used." The standard of identify goes on to further describe that "parmesan cheese, reggiano cheese" is prepared from milk (which may be pasteurized or clarified or both, and may be bleached with benzoyl peroxide) and other ingredients that include artificial coloring, and enzymes of animal or plant origin that aid curing. It is cured for not less than 10 months.[5]

It is ironic that shortly after the 2003 GI hearings, Kraft petitioned the FDA to amend standards for Parmesan and reggiano cheese. In a December 23, 2005, letter in response to Docket No. 2000P-1491, Kraft's petition sought to amend the standard of identity to reduce the curing time for Parmesan to 6 months. It wrote that, in 1997, it had determined that modern enzyme technologies allow Parmesan to "be of the same basic nature and essential characteristics" of Parmesan aged for a longer time. It also cited that length of curing time is not a safety factor, as Kraft's own expert microbiologists "know that within the first two to three weeks of the aging process, undesirable microorganisms—assuming that any are produced—become unable to compete with the starter culture. In addition, the low moisture and relatively high salt content of parmesan cheese contribute to its good safety profile." The company cited its August 2005 comments as agreeing with the prevailing food industry position to modernize food standards, "that a horizontal approach can make the best use of limited agency resources." In this letter, Sheryl Marcouiller, Chief Counsel, Food Law from Kraft stated, "Food standards should permit maximum flexibility in the technology used to prepare a standardized food, so long as the technology doesn't affect the food's basic nature, essential characteristics, nutritional quality or safety." This statement provides stark evidence of the divide between European, consumer-driven values and the economics driving food production in the United States. Marcouiller further wrote that:

> A horizontal approach could be used to increase flexibility for parmesan and other cheeses by establishing an across-the-board

> authorization for shorter cure times made possible by advances in technology, such as enzyme technologies that allow for faster ripening. Alternatively, FDA might consider amending the parmesan cheese standard to provide for any aging period resulting in a cheese with the basic nature and essential characteristics as parmesan, as determined through appropriate testing and other data.

It is no wonder the EU sought protection of its traditional cheeses.

In contrast, in the EC, the name Parmigiano Reggiano reflects a place of production: *Parmigiano* is the Italian adjective for Parma, a city in northern Italy, in the region of Emilia-Romagna. The English translation of *Reggiano* is "from Reggio Emilia," another city in northern Italy. European law classifies the name, Parmigiano Reggiano, as well as the translation "Parmesan," as a protected designation of origin. Other governments have agreed; in England, when you buy Parmesan, you get Parmigiano Reggiano. The PDO specifies that Parmigiano Reggiano is a hard cheese made from raw cow's milk partially skimmed by natural surface skimming.[6] The milk must not undergo any heat treatment and must come from cows fed primarily (75 percent) on fodder obtained in the area of origin, with silage feeding prohibited. It must be produced exclusively in the Italian provinces of Parma, Reggio Emilia, Modena, and also in parts of the provinces of Mantua and Bologna. It must be made from raw cow's milk. Milk is obtained from both evening and morning milkings, with the milk from the morning milking placed in copper vats and mixed with the evening milk. Native whey is added, and the use of starters is not permitted. Following coagulation, curd is broken and cooked, drained and molded to form classic wheels. The wheels are brined, marked, then aged for at least 12 months.

The links to the geographic region are specified. The soil characteristics of the land in combination with the climate conditions of the geographical region influence the composition of the natural microflora of the feed. Faithfully handed-down traditional practices, established over many centuries, are still in use today in the area of production. The physical, chemical, and microbiological properties of the milk are due to the manner in which the cows are fed and ensure the typical character and quality of Parmigiano Reggiano cheese. The cheese maturation must be performed within the defined geographical region: "By virtue of its specific climatic conditions [it] is a necessary phase in order to ensure that the product obtained from the processing of milk can acquire, through particular enzyme processes, the characteristics proper to a 'Parmigiano Reggiano' cheese."

Only the whole cheese bearing the protected designation of origin may be grated; it must be packaged immediately afterward without any processing or addition of substances likely to modify its conservation properties or original organoleptic characteristics. While traceability rules are a new concept in modern food safety, the Parmigiano Reggiano consortium was there many years before. "The identification marks on each wheel comprise the words Parmigiano Reggiano next to the registration number of the dairy and the year and the month of production stenciled onto the surface of the heel, the oval stamp with the words '*Parmigiano Reggiano Consorzio Tutela*' and a casein nameplate showing the codes identifying the mold and where appropriate, the mark identifying a second class cheese."[7]

As Larry Olmsted, author of *Real Food, Fake Food*, noted in his column in *Forbes*:

> I noted in my last column that by law, Parmigiano-Reggiano is allowed to contain only three very simple ingredients: milk (produced in the Parma/Reggio region and less than 20 hours from cow to cheese), salt, and rennet (a natural enzyme from calf intestine). Three other ingredients, Cellulose Powder, Potassium Sorbate, and Cheese Cultures are not found in Parmigiano-Reggiano—they are completely illegal in its production. Yet all three are in Kraft 100% Grated Parmesan Cheese (I'm not sure if that means it is supposed to be 100% "parmesan" or simply 100% grated, which it certainly is). It's far enough from the real thing that Kraft was legally forced to stop selling its cheese labeled Parmesan in Europe.[8]

I am not sure that I support the entire US GI argument. A case in point is Sartori, a Wisconsin cheese firm that produces BellaVitano cheese. What is BellaVitano? As stated on the Sartori website: "A Sartori-family original, this rich, creamy cheese with its nutty, fruity flavor is also—in all modesty—a celebrated gold-medal winner. Inspired by traditional, Italian farmstead cheese, BellaVitano Gold begins in the mouth like a premium Parmesan and finishes in award-winning style with hints of melted butter. This is where our artisan cheesemaking first began to shine." Sartori does not need to be threatened by GIs. It produces a wonderful cheese that speaks for itself in terms of character, quality, flavor, complexity, and richness, regardless of the name of the cheese style. I am a big fan of this cheese. US artisan producers have largely avoided encroachment on the GI issue. Successfully marketed products carry the names Wabash Cannonball, Red Hawk, Hooligan, Harbison, Barely Buzzed, Tarentaise, Cremont, and Pleasant Ridge Reserve, to name a few. We are producing some fabulous cheeses here in the United States that can compete with the best cheeses produced anywhere in the world. As evidence, the *Washington Post* ran a story in 2017 by Jason Wilson titled "How good has U.S. cheese become? Good enough to worry the Italians."[9]

On October 16, 2003, the Consortium of Parmigiano Reggiano filed a complaint with the European Court of Justice concerning the improper use by certain German companies of the name Parmesan. In the complaint, the consortium indicated that these products neither exhibited the properties of Parmigiano Reggiano nor came from the region. On February 26, 2009, the European Court of Justice ruled in favor of the consortium, stating that a GI could become a generic name over time but only cheeses bearing PDO Parmigiano Reggiano can be sold as Parmesan.[10]

In 2003, the United States filed an update to the 1999 WTO dispute over the EC's geographic protections, and a parallel dispute was filed by Australia. The dispute arose over whether Council Regulation (EEC) No. 2081/92 on GIs and Designations of Origin conflicted with trademark protections under the Agreement on Trade-Related Aspects of Intellectual Property Rights (TRIPS). Australia and the United States raised concerns over the regulation's treatment of trademark registrations that were in conflict with later-registered GIs.

A Dispute Settlement Body (DSB) was convened by the WTO. In March 2005, the DSB issued a decision regarding the conflict between trademarks and geographical indications. The WTO panel was in general agreement with the Australian and US positions, but ruled that, under certain circumstances, exceptions applied, allowing the coexistence of GIs and prior trademarks. This ruling resulted from the long-standing complaint by the

United States that the EC's GI system discriminates against foreign products and persons—primarily by requiring that EC trading partners adopt the EC-style system of GI protection. In addition, the United States argued that the EC's GI system provides inadequate protections to trademark holders. In the report adopted by the DSB, the WTO panel agreed that the EC's GI regulation discriminates against non-EC products and persons. The panel also agreed with the United States that Europe could not, consistent with WTO rules, deny US trademark owners their rights; it found that, under the regulation, any exceptions to trademark rights for the use of registered GIs were narrow, and limited to the actual GI name as registered. Otherwise, there was no finding that the substance of the EC system of GI protection, which requires product inspection, was inconsistent with WTO obligations. The panel recommended that the EC amend its GI regulation to come into compliance with its WTO obligations. The EC, the United States, and Australia agreed that the EC would have until April 3, 2006, to implement the recommendations and rulings."[11] At the DSB's April 21, 2006, meeting, the EC indicated that it had fully implemented the DSB's recommendations and rulings by adopting a new regulation that came into force in March 2006. However, Australia and the United States disagreed that the DSB recommendations had been fully implemented and invited the European Communities to take note of their comments in making further revision of the newly promulgated regulation.

In 2012, the EU, recognizing that it needed to address some of the issues raised by the WTO, as well as to clarify and simplify other rules, implemented Regulation (EU) No. 1151/2012 of the European Parliament and of the Council of November 21, 2012, on quality schemes for agricultural products and foodstuffs.[12] The 2012 regulation repealed Council Regulations (EC) No. 510/2006 of March 2006 on the protection of geographical indications and designations of origin for agricultural products and foodstuffs, and it repealed Council Regulations (EC) No. 509/2006 on agricultural products and foodstuffs as traditional specialities guaranteed.[13] The regulation improved and updated the framework for the protection and promotion of quality agricultural products. In particular, the scope of the protection was extended to new products, some definitions were aligned to TRIPS, the protection was enhanced, registration and amendment procedures were streamlined, and the use of the symbols for Protected Designation of Origin, Protected Geographical Indication, and Traditional Speciality Guaranteed (TSG) became mandatory for products of EU origin. The TSG scheme was strengthened, the rules on controls were clarified, and a scheme for optional quality terms was established.

K. William Watson of the Cato Institute offered advice to US trade negotiators on how to best deal with what he calls "the reign of terroir": "It may be tempting for U.S. T-TIP negotiators to follow the model used in the Canada-EU Comprehensive Economic and Trade Agreement (CETA). In CETA, Canada accepted an obligation to protect the EU's list of GIs, but with limitations on some of the more contentious names. In particular, five generic cheese names—Asiago, feta, fontina, gorgonzola, and muenster—are still allowed "when the use of such terms is accompanied by expressions such as 'kind,' 'type,' 'style,' 'imitation,' or the like, and is in combination with a legible and visible indication of the geographical origin of the product concerned. Importantly, Canada can also continue to use the term Parmesan."[14]

To this day, the protection of common food names remains a point of contention between EC and US cheese producers. Could this be the reason that FDA amended its *E. coli* regulations in 2009 and 2010, and later proposed a ban on use of wood in cheese aging? As evidence that this continues to be an issue, the Consortium for Common Food Names (CCFN) was formed in 2012 to protect the right of food producers to use common food names. Founding members of this consortium include the US Dairy Export Council (USDEC) and several leading US cheese companies.[15]

Mexico is the largest export market for cheese produced in the United States. On April 27, 2018, the EU and Mexico reached a new trade agreement that will provide the EU with greater access to Mexico's dairy market. Notably, in the intellectual property area, the agreement provides protection of 340 GI products, including cheeses. The US response? Tom Vilsack, the former Secretary of Agriculture and now president and CEO of USDEC stated:[16] "We are deeply disappointed that Mexico has limited US access by restricting the use of common food names that have been used in the Mexican market for years. This undermines the rule of law and the value of the market access terms the US has long had in place with Mexico." Generic terms such as Parmesan, feta, muenster, gorgonzola, Asiago, fontina, and Neufchâtel will have future restrictions despite the fact that these names have been previously used in Mexico.

Handler wrote that in response to the EU's attempts to make its model of GI protection a global standard, the United States will likely settle future GI disputes by imposing requirements on its trading partners.[17] A case in point is the new United States-Mexico-Canada Agreement (USMCA), a signed but not yet ratified agreement meant to supercede the North American Free Trade Agreement (NAFTA). On October 8, 2018, the USMCA included language regarding GI policy and the protection of common food names. Under this agreement commonly used cheese names that may not be restricted by Mexico moving forward include *mozzarella*, *cheddar*, *provolone*, and others. In addition, Canada and Mexico will be adopting GI parameters that make it more difficult for any nation to register new GIs that are common food names, and common name users will be able to oppose GI applications that would monopolize use of generic terms. While praising US government leadership in safeguarding generic terms, the CCFN expressed its disappointment in the Mexican support for the EU position on protection of common food names.

The CCFN stated that "CCFN's work in the region is not yet done. . . . USMCA marks a sea change in GI policy, recognizing the equal importance of the protection of distinctive products through GIs, and the defense of generic terms long used in the marketplace," said Jaime Castaneda, executive director of CCFN. "The U.S. Administration demonstrated great leadership in pushing forward many key concepts for effective GI policy, which are of benefit to consumers and producers throughout North America, and which CCFN has long promoted and worked on with government leaders. These include commitments on transparency and the ability for stakeholders to object to pending GIs that may infringe on their rights to use generic terms."[18]

The USMCA marked the first time the United States has specifically included reference to the rights of generic name users within a trade agreement—a goal that the CCFN had been working toward for many years. "These explicit considerations safeguarding generic terms are essential," said CCFN Chairman Errico Auricchio, president of BelGioioso Cheese, "because the EU continues to move the lines on which names of cheeses, meats, wines and other products are fair game when it comes to abusing GI policies and monopolizing common names and terms."[19]

These activities become interesting if we place them within the context of our timeline. In July 2015, the FDA's Dr. Susan Mayne spoke in a highly anticipated session at the American Cheese Society Annual Conference in Providence, Rhode Island. She was newly appointed in her position at FDA as the director of CFSAN (Center for Food Safety and Applied Nutrition) and pledged to work closely with the American Cheese Society. It was curious that her session was sponsored by the Consortium for Common Food Names.

As to the March 11, 2014, letters from members of the US Senate to Secretary Vilsack, a March 14, 2014, article by EURACTIV.com with Reuters titled "US Senators shocked by EU's cheese-name claims" stated: "The senators said their action was supported by Kraft Foods Group, Denver-based Leprino Foods, the world's largest mozzarella maker, and groups such as the National Milk Producers Association, US Dairy Export Council, and the American Farm Bureau Federation."[20]

John Sheehan, director of FDA-CFSAN's Division of Dairy, Egg, and Meat Products, and Monica Metz, chief for FDA-CFSAN's Milk and Milk Products Branch, were former employees of Leprino Foods.[21]

CHAPTER 11

The Food Safety Modernization Act

Previous chapters in this book make reference to the Food Safety Modernization Act, or FSMA. What exactly is the FSMA, and how did it become law? And what are the implications of the FSMA to the artisan cheese industry, both here in the United States and abroad? The Food Safety Modernization Act was signed into law in 2011. This was the most sweeping reform of US food safety law since the Federal Food, Drug, and Cosmetic Act (FFDCA) was promulgated back in 1938. This new law provides the FDA with unprecedented regulatory power and authority. In order to understand why the FSMA was promulgated, it is important to understand the backdrop for the FSMA. While legislative efforts began in earnest in 2007, there had been discussions many years beforehand of the need to reorganize and streamline the efforts of the disparate federal agencies having food safety oversight within a single food safety agency that would oversee the safety of US-produced foods. Critics argued that, in the United States, $1 billion was being spent annually by the 12 different federal agencies that administered the 35 laws that constitute the Federal Food Safety System. Beyond the federal government, state and local government regulations are superimposed upon federal regulations. In spite of this vast effort, the best estimates from the Centers for Disease Control and Prevention (CDC) in 2011 indicate that we still have 48 million cases of foodborne illness each year in the United States that result in 128,000 hospitalizations and 3,000 deaths, and, sadly, most of these illnesses and deaths are fully preventable.[1]

The years between 2006 and 2007 were not good ones for food safety in the United States. An outbreak that infected 628 people in 47 states was linked to peanut butter.[2] Spinach was linked to an outbreak of *E. coli* O157:H7 that caused three deaths and sickened 205 individuals.[3] *Listeria* in pasteurized milk killed two people in Boston.[4] A total of 21.7 million pounds of frozen hamburger was recalled following reports of 25 illnesses in eight states from *E. coli* O157:H7. The CDC reported that in 2007 alone, 1,097 confirmed outbreaks occurred, which resulted in 21,244 reported cases of illness.[5] In addition to human illness, the FDA received 17,000 consumer complaints about pet food that sickened dogs and cats because it contained ingredients from China contaminated with the industrial chemical melamine.

In response to the public outcry over these incidents, H.R. 2749, the Food Safety Enhancement Act, was introduced in 2009.[6] The purpose of this bill was to give the FDA greater regulatory powers over the national food supply and food providers, with the goal of preventing foodborne illnesses instead of reacting to them when they occur, thereby ensuring food safety. More specifically, it would increase the frequency of FDA inspections of food processing plants, expand the FDA's traceback capabilities when outbreaks occur, give the FDA mandatory recall authority, and require food facilities to have safety plans in place in order to mitigate hazards. The bill was an enhancement to previously proposed food safety bills introduced in the 111th Congress.

Senator Dick Durbin was the sponsor of a Senate version of the bill, S. 510, the FDA Food Safety Modernization Act of 2009.[7] A conference committee combined provisions of S. 510 and H.R. 2749. In introducing the legislation, Senator Durbin cited the recalls of tainted peanut butter, spinach, seafood, and pet food as evidence of our nation's broken food safety system. The bill noted that authority for regulating our nation's food supply was split among more than 10 federal agencies, and most lacked the resources that were necessary to ensure the integrity of US consumed foods. Senator Durbin stated:

> I am working to secure increased resources for the Food and Drug Administration (FDA) food safety program, and I am the lead sponsor of several pieces of legislation designed to improve our food safety system. Among these bills is the Safe

> Food Act, which would streamline our food safety structure, improving coordination by combining the disparate food safety functions spread across the federal government into a single agency based on scientific principles. We also need to establish a robust system for overseeing the safety of imported food. I introduced legislation to strengthen the FDA's ability to monitor and inspect goods that enter the U.S. from abroad by imposing a fee on companies and countries exporting food products to the U.S. Consumers often take for granted that the food they purchase will be safe whether it originated in the U.S. or was imported, but the standards of other countries and the lack of U.S. food inspectors monitoring imported food too often proves them wrong.[8]

The FSMA amended the Federal Food, Drug, and Cosmetic Act (FFDCA) by expanding the authority of the Secretary of Health and Human Services (HHS) to regulate food, including authorization for the HHS secretary to suspend the registration of a food facility. Among the sweeping reforms, the act required each food facility to evaluate hazards and implement preventive controls. The voluntary qualified importer program required the HHS and the US Department of Agriculture (USDA) secretaries to prepare the National Agriculture and Food Defense Strategy (NAFDS). The HHS secretary is also required to identify preventive programs and practices to:

- Promote the safety and security of food
- Promulgate regulations on sanitary food transportation practices
- Develop a policy to manage the risk of food allergy and anaphylaxis in schools and early childhood education programs
- Allocate inspection resources based on the risk profile of food facilities or food
- Recognize bodies that accredit food-testing laboratories
- Improve the capacity of the secretary to track and trace raw agricultural commodities

The act also required the HHS secretary, acting through the CDC, to enhance foodborne-illness surveillance systems. For the first time in history, the act authorized the secretary to order a food recall. Up to this point, guns, cars, toasters, and toys could be recalled, but not food. On the import side, the act provided for foreign supplier verification activities, a voluntary qualified importer program, and the inspection of foreign facilities registered to export food.

While some US consumer advocates were cheering the legislation, our neighbors to the north were not so gleeful. The *Canada Free Press* observed, "As it is currently written, the bill, S. 510, will actually make our food less safe. S. 510 will strengthen the forces that have led to the consolidation of our food supply in the hands of a few industrial food producers, while harming small producers who give consumers the choice to buy fresh, healthy, local foods." The Organic Consumers Association agreed, writing: "Although reform of the industrial food supply is clearly needed, this bill threatens to create more problems than it will solve. S. 510 would undermine the rapidly growing local foods movement by imposing unnecessary, burdensome regulations on small farms and food processors—everyone from your local CSA to the small bakers, jam makers, and people making fermented vegetables to sell at the local farmers market."[9] The other obvious problem was that the FSMA applied only to FDA-inspected foods. Meat and poultry products that fall under mandated inspection from the USDA Food Safety and Inspection Service (FSIS) do not come under the FSMA regulation. Therefore, despite the original intent of Congress to create a single food safety agency, the FSMA did not achieve the goal of organizing food safety under the umbrella of a single federal agency.

The US Senate Committee on HELP (Health, Education, Labor, and Pensions) unanimously passed S. 510, and on January 4, 2011, President Obama signed the bill into law. The Food Safety Modernization Act changed the focus on food safety from reaction to prevention. In order to prevent outbreaks of foodborne illness, the law mandates that every food producer must have a written food safety plan based on hazard analysis and risk-based preventive controls, and this plan must be overseen by a qualified individual. The law also provided, for the first time in history, mandatory FDA inspections of foreign facilities that export food to the United States. The FSMA gives the FDA authority to deny entry of food from a foreign facility if the FDA is denied access to the facility or country where the food is produced. The foreign facility producing food must conform to

US standards. How this requirement affects the production of traditional foods abroad, such as cheeses, remains to be seen, but impacts are already being felt. The law expands the authority of the FDA to inspect foreign food facilities, including factories and warehouses. It states that the FDA or its designee should be permitted to enter a facility within 24 hours (or within an agreed-upon time), and that food from the facility will not be allowed into the United States if this access is postponed or denied.[10]

Ironically, one of the original goals of the legislation that ultimately became the FSMA was the harmonization of our food safety standards with those of our international trading partners.[11] As noted in earlier chapters, at least on the cheese front, the US *E. coli* standards are inconsistent with the EU approach to ensuring food safety. I would argue that the EU has the approach that best protects consumers and ensures the safety of cheeses. But that issue aside, could the FSMA now be used to conform European cheese production to US standards? And if so, what are the implications?

To implement the FSMA, the FDA drafted, and has now finalized, seven framework rules. FDA was under pressure from Congress to have the framework rules finalized in a timely fashion, and some of the rules as proposed, particularly the Produce Rule, created much tension between the FDA and the US produce industry. Other rules, such as preventive controls for human food, sought to "modernize" Good Manufacturing Practices. The old regulations were part of 21 CFR Part 110, a regulation that specified Good Manufacturing Practices. The new rules under 21 CFR 117 subpart B ("Current Good Manufacturing Practice, Hazard Analysis, and Risk-Based Preventive Controls for Human Food") might seem familiar to many, as they use a pharmaceutical template to align them more closely with 21 CFR Part 210 ("Current Good Manufacturing Practice in Manufacturing, Processing, Packing, or Holding of Drugs"). However, foods are not pharmaceuticals, and controls designed to ensure sterility in the drug manufacturing process are inappropriate for applications to foods. We do not want foods, particularly products such as cheese and produce, to be *sterile*; we need them to be *safe*. We are already seeing recalls of products long assumed to be low-risk, such as flour, as a source of *Salmonella*. The system in place to transport grain and mill flour was never designed to be a sterile process, as flour is not a ready-to-eat product. It is assumed that it will be cooked before consumption. But under the FSMA, and through guidance from FDA, the lines between ready-to-eat foods and not-ready-to-eat foods (those requiring further processing, such as cooking), are becoming blurred. In the case of flour, outbreaks have occurred because consumers ate raw cookie dough.

American consumers and food producers should pay close attention to the FSMA. If the FDA approaches the produce industry in the same manner as it has chosen to "regulate" the artisan cheese industry, the FSMA will fail miserably and cause significant disruptions in the ability of produce growers to sell and market their products, without assuring safety. As a case in point, in 2018, the entire US romaine lettuce industry was shut down due to an outbreak of *E. coli* O157:H7 illness that affected 62 individuals.[12] The outbreak strain matched a strain of *E. coli* O157:H7 from a water reservoir that may have come into contact with lettuce produced on a farm in Santa Barbara County, California. During the outbreak, and before the suspected source of the lettuce was identified, the CDC advised consumers nationwide to refrain from eating romaine lettuce. The source of the contaminated lettuce could not be definitively identified, and, as a result, the entire US romaine lettuce industry paid the price. Unfortunately, despite legislative efforts going back to 2007, as well as the signing of the FSMA into law, along with produce rules that have gone into effect, we still are experiencing outbreaks that cause massive disruption and economic consequences for our food supply. While public health must be protected, so, too, must our industries that produce the food we eat. The economic consequences to the romaine lettuce industry nationwide are staggering. Does the FDA have the knowledge base and trained inspectors needed to improve produce safety?

Large-scale produce outbreaks are not new to US consumers. These well-publicized outbreaks caused many consumers to ask fundamental questions such as how much do they know about where their food comes from, and under what conditions is it produced? As discussed previously, an outbreak of foodborne illness linked to spinach contaminated by *E. coli* O157:H7 occurred during the early fall of 2006. When the final outbreak tally was over, 205 cases of illness and 3 deaths resulted. Federal and state officials could not pinpoint the exact source of *E. coli* O157:H7 and how and why spinach became contaminated, but several things became clear. During the outbreak investigation, CDC, FDA, and California officials found the outbreak strain in 13 bags of Dole brand baby spinach produced during an August 15, 2006, production run at a facility in San Juan Bautista, California.

However, no samples matching the outbreak strain were found during environmental testing of the production facility. The product code was traced to four fields in Monterey and San Benito Counties. While *E. coli* O157:H7 isolates were found in environmental samples on all the farms, samples that matched the outbreak were limited to one farm in San Benito County, where the outbreak strain was isolated from river water, along with cattle and wild-pig feces. Less than a mile from the spinach field was a grassfed cattle operation. Evidence of wild pigs in and around the cattle, spinach-growing, and irrigation-well areas was found during the investigation.[13]

First, this outbreak clearly illustrates the dangers associated with massive-scale centralization of food production. The spinach in question was grown in the three California counties that supply spinach to most of the major companies distributing spinach nationwide. Outbreaks are rare and exceptional events; the majority of spinach grown in this area and consumed nationwide is done so safely and without incident. Second, we have excellent surveillance systems operational in the United States that tell us when things are going wrong. In 1996, CDC implemented a system known as PulseNet, which tracks DNA "fingerprints" of bacterial isolates reported from state public health labs around the country.[14] Public health officials can know quickly when a multistate outbreak is happening because all state public health labs are networked with CDC and can quickly compare DNA data on outbreak strains common to several states. PulseNet transformed foodborne disease surveillance from a passive system, where most outbreaks went unrecorded, to an active system designed to rapidly remove contaminated foods from commerce. This system is saving lives and is certainly working to facilitate removal of contaminated foods from commerce. Under the FSMA, the US surveillance system has been further improved with GenomeTrakr, which brings yet more sophistication and certainty to outbreak investigations, because the DNA of the entire genome of an organism can be sequenced and definitive proof of a link to contaminated foods can be provided.[15]

Unfortunately, routine testing and quality assurance conducted by some food companies have not kept pace with this fundamental public health change. State and federal regulatory agencies are actively using scientific tools to precisely identify foodborne pathogens in foods that match genetic fingerprints from patients who have consumed contaminated products.

Finally, given the severity of consequences associated with *E. coli* O157:H7 illness, there is a need to alert consumers. Just a few cells of this particular pathogen can permanently inactivate kidney function in young children. This particular illness, hemolytic uremic syndrome, is very serious.

There are several theories as to how the spinach from the 2006 outbreak became contaminated, including changes in groundwater levels during the growing season that could have contributed to contamination.[16] One theory is that surface river water flowed into the valley where the Paicines Ranch is located, percolated into the ground, and recharged the groundwater. Alternatively, the contamination could have resulted from flooding events, which create a perfect storm for contamination. Another theory is that the contamination originated from fertilizer used on the spinach field, which consisted of heat-treated chicken manure pellets. These theories afford insight into the complexity of the spinach outbreak investigation. The HHS said that although the investigation was not able to definitively determine how the *E. coli* got onto the spinach, it was the first time that a clear link between an individual who became ill from a contaminated product and a farm source had been made down to the farm level. Many of the potential disease vectors investigated during this outbreak—animals in fields, irrigation water, and fertilizer—became identified as hazards that farmers now had to control under the proposed FSMA produce rule.

The 2018 romaine lettuce outbreak, and the CDC and FDA's responses to warn consumers nationwide to avoid consuming romaine lettuce, thereby shutting down the entire industry, was eerily familiar. In 2008, an outbreak involving *Salmonella* serotype Saintpaul occurred between April 1 and September 4, 2008. Over 1,500 cases of illness and 2 deaths resulted from this outbreak.[17] On May 22, 2008, the New Mexico Department of Health had notified the CDC about 19 cases of *Salmonella* infection with a common PFGE pattern, a DNA fingerprint that determines bacterial relatedness. The epidemiological investigation conducted by the CDC had identified eating raw tomatoes as being strongly associated with illness, and on June 7, the FDA and CDC issued an advisory, warning consumers nationwide to avoid eating tomatoes. As a result of this warning, the entire US tomato industry was impacted. However, tomatoes were not the source of the outbreak. Despite the CDC's continued warnings against tomato consumption, cases of *Salmonella* serotype Saintpaul infection continued to increase, leading the CDC to conduct additional epidemiological studies. Eating at a Mexican-style restaurant; eating pico de gallo salsa, corn tortillas, or salsa; and having raw jalapeño peppers in the household were all found to be factors associated with illness. Subsequent field investigations recovered the outbreak strain of *Salmonella* from jalapeño peppers collected in Texas, as well as agricultural water and serrano peppers on a Mexican farm. Jalapeño and serrano peppers had been previously identified as a major vehicle for *Salmonella* transmission, yet FDA officials said they were surprised by the outbreak because Mexican peppers had not been spotted as a problem. However, the FDA's own records showed that peppers and chilies were the top Mexican crops rejected by border inspectors. This study proves that epidemiology can be wrong, and serves as a cautionary tale for those in the produce industry who were economically affected. FDA and CDC officials noted that if tomatoes could have been traced from a retail store or restaurant back to the source within 48 hours of being implicated, investigators would have realized tomatoes were not the likely source. The Pew Charitable Trusts estimates the cost to the Florida tomato industry alone was $100 million.[18]

Traceability rules have become part of the FSMA. But there is a fundamental problem: Is the FDA ready to assume larger authority over produce safety, an area where it admittedly lacks expertise? Of inspection of foreign and domestic artisan cheese production, where the FDA has admitted it lacks trained and knowledgeable inspectors? The French government spends vast resources on cheesemaker education and training. Do FDA inspectors have equivalent credentials and knowledge?

Back in 2008, what did the FDA propose as a solution to the lettuce and spinach outbreaks? The FDA fast-tracked approval for use of irradiation on these products. The advocacy organization Food & Water Watch, critical of the FDA's approach, stated:

> Today, the U.S. Food and Drug Administration announced it will allow fresh spinach and iceberg lettuce to be treated with ionizing radiation. Nearly two years after a major *E. coli* outbreak was linked to California spinach, it is unbelievable that the FDA's first action on this issue is to turn to irradiation rather than focus on how to prevent contamination of these

> crops. This just illustrates once again how misplaced this agency's priorities really are. Instead of beefing up its capacity to inspect food facilities or test food for contamination, all the FDA has to offer consumers is an impractical, ineffective and very expensive gimmick like irradiation. . . . Irradiation is a Band-aid, not a cure. Allowing spinach and lettuce to be irradiated would simply mask unsafe production practices, while supplying lower quality, less nutritious and potentially hazardous food. Instead of pursuing irradiation, vegetable growers and processors should improve flawed sanitation practices and FDA should inspect vegetable-processing plants more thoroughly. American consumers expect more and deserve better than questionable treatments like irradiation.[19]

Will these be the tools used in the future by the FDA to ensure produce safety, and if so, will consumers have a voice in this decision-making? For too long, the FDA and other US public health agencies have focused on technologies to achieve the standard 5-log kill of pathogens to ensure food safety. With an approach such as irradiation, where is the incentive to produce foods with the highest-quality ingredients? Advertently, or perhaps inadvertently, the FDA is sending a chilling message to consumers.

Inspection of Food Facilities

As the FDA was busy swabbing artisan cheese facilities to identify sources of *Listeria* contamination, the Office of Inspector General (OIG) of the HHS was busy issuing a report: "FDA Inspection of Domestic Food Facilities, April 2010." The OIG for the HHS is charged with identifying and combating waste, fraud, and abuse in the more than 300 programs administered by the HHS, including Medicare and programs conducted by agencies within HHS, such as the FDA. The Senate Committee on Agriculture, Nutrition, and Forestry requested that the OIG review the extent to which the FDA conducts food facility inspections and identifies violations. The objectives of the OIG report were, therefore, to determine the extent to which the FDA conducts inspection of domestic food facilities, identifies violations in food facilities, takes action against them, and ensures that these violations are corrected. Using the FDA's own records from 2004 to 2008, the OIG found that the FDA inspected less than 25 percent of domestic food facilities each year.[20] The OIG also shockingly found that 56 percent of all US food facilities had gone five or more years without an FDA inspection. The FDA took action against 46 percent of facilities with official action indicated (OAI) classifications (indicating serious situations that could impact consumer health and safety), but for the remainder, the FDA either lowered the classification or took no action. For 36 percent of facilities with OAI classifications in FY (fiscal year) 2007, the FDA took no additional steps to ensure that violations were corrected. As significant weakness was found with the FDA's domestic inspections program, the OIG recommended increasing the frequency of facility inspections, with emphasis on high-risk facilities; providing guidance on when it is appropriate to lower OAI classifications; taking appropriate action against OAI facilities, especially those with a history of violations; ensuring that violations are corrected at OAI facilities; and considering statutory authority to impose civil penalties against noncomplying facilities, and authority to access records during inspection.

In 2017, 10 years following the initial OIG report, the OIG conducted a follow-up review to determine if the FDA had made progress with respect to its inspection of domestic food facilities. The OIG looked at data from the FDA inspections conducted between 2010 and 2015. The OIG concluded that challenges remained with respect to the FDA's ability to conduct facility inspection. Some of the issues revealed in the report included a decrease in the overall number of food facilities that the FDA inspected since the passage of the FSMA, from a high of 19,000 facilities in 2011 to just 16,000 facilities in 2015. As found in the preceding report, the FDA did not always take action when it found OAI violations. When action was taken, facilities were left to voluntarily correct the violations. The FDA was found to rarely take advantage of new administrative tools that the FSMA provided, such as eased criteria for administrative detention of potentially unsafe food, use of recall authority, and suspension of registration of a facility to stop that facility from distributing food. The FDA's actions were not always timely and did not always result in the correction of these violations. The FDA consistently failed to conduct timely follow-up inspections to ensure that significant inspection violations were corrected. For approximately half of the inspection violations that were found to be significant, a follow-up inspection was not conducted for a year, and for 17 percent of the significant inspections violations, no follow-up inspection was conducted at all.[21]

The OIG report highlights an example of one of the violations: "FDA found unsanitary conditions in a New Mexico facility that manufactures chili peppers and spices. FDA investigators took environmental samples at the facility, and 21 came back positive for *Salmonella*. The strain of *Salmonella* uncovered in this inspection was identical to the strain discovered in the plant during a previous inspection." During the monthlong late-December 2018–January 2019 government shutdown, it became ironic to listen to FDA officials warn the public that they were at risk because the FDA could not do food inspections. The reality is that, for a long time, the FDA has not conducted timely inspections of food processing facilities, and this became of concern to members of Congress. It is precisely why the mandated inspection goals became part of the FSMA. Of further important note is that at no time during the government shutdown was meat inspection affected. It was food safety business as usual at the USDA-FSIS.

The Produce Safety Rule

With the FSMA, the FDA was provided with authority for food safety oversight in new areas, such as farms. Historically, the USDA is the agency that has conducted agricultural research, and most of our federal knowledge and expertise with respect to farming and agricultural production practices resides at the USDA Agricultural Research Service (ARS) in Beltsville, Maryland, and through the larger USDA Cooperative State Research Services (CSRS) that provides funding to land grant colleges and universities throughout the United States. When the FDA was writing the Produce Safety Rule, it could have easily engaged in dialogue with experts at ARS, or included them as partners in drafting the regulations. But it did not, and as a result, what was rolled out by the FDA in the initial draft of the Produce Safety Rule became of great concern to not only produce growers but also individual State Department of Agriculture officials across the United States. Some of the problematic regulations proposed in the draft Produce Safety Rule included a 270-day (nine-month) mandatory withholding period from the time of manure application to fields until edible crops from those fields could be harvested. Growers in my home state of Vermont were enraged; they have one of the shortest growing seasons in the United States, and now the FDA was proposing to effectively further reduce this already narrow growing window. Where was the science that supported this proposed rule? Growers were also confused: The USDA's National Organic Program standards stipulated a 180-day withholding period from the time of manure application until edible crop harvest. Why didn't the FDA propose rules that were consistent with the USDA National Organic Program standards? In fact, it became evident that the "extensive" science behind the FDA's proposed withholding requirement consisted of results of a single study conducted at one field site in the United States. The research had been conducted by USDA-ARS in Beltsville, using soils amended with poultry manure. The type of soils, type of manure, and climate conditions are extremely variable across the United States. Why would the FDA base a rule with such sweeping implications across the United States on such limited scientific data?

The water standards proposed by the FDA in the Produce Safety Rule were equally contentious. The United Fresh Produce Association (UFPA) offered some of the best feedback to the FDA in its public comments on the FDA's draft Produce Safety Rule. The UFPA suggested that since the FDA "had not previously inspected farms, due to the FDA's lack of knowledge of farming activities, the FDA staff will require training or risk alienating the industry. Without an adequate understanding of industry practices, inspectors will risk rejecting workable best practices due to lack of knowledge. Healthy interactions between regulated and regulator can help lead to voluntary compliance; this is the only way food safety will be advanced, as there is not enough funding to enforce our way to compliance." Based upon my experience with the FDA's oversight of artisan cheesemaking practices, the artisan cheesemaking community would heartily support the UFPA's assessment.

The UFPA stated to the FDA that, in its view, it was compliance with water quality standards promulgated in the proposed rule that presented the single greatest obstacle for fruit and vegetable producers regulated under the Produce Safety Rule. Many farms use surface water, and the UFPA indicated that the water quality requirements and water testing intervals proposed in the rule would put farms out of business. In the Produce Safety Rule, instead of recognizing the global standards of <1000 cfu/100 ml fecal coliforms established by WHO as appropriate for irrigation water, the FDA used the EPA recreational water standard of 235/100 ml generic *E. coli.*[22] As stated by the UFPA, "FDA provided no scientific rationale or justification for this testing beyond a single commodity and only on a single region's data. We believe that current science is inadequate to justify a fixed test organism, number or testing requirement."

Additional, and fortunately, very useful comments on the Produce Safety Rule were offered by the National Association of State Departments of Agriculture (NASDA). The NASDA is a nonpartisan association that represents the commissioners, secretaries, and directors of the departments of agriculture in all US states. NASDA members cautioned the FDA that the proposed produce rule was a new area of regulation for the federal government. Previously, the FDA conducted investigations on farms only when produce was suspected of causing a foodborne illness outbreak, or in cases where a farm qualified as a food processing facility. Under the proposed regulations, the NASDA cautioned:

> Farmers will be subject to inspections as a matter of routine. This new area of responsibility will take considerable thought, training, education and understanding by all parties. Regardless of existing authority, the individual state departments of agriculture should have a significant role in implementing the new requirements that affect producers because of their valuable relationships and insight. As a result of the 2013 FDA cantaloupe assignment, it has become clear to NASDA and NASDA members that the FDA is neither prepared nor equipped to handle on-farm inspections of produce operations or the outreach necessary to bring farmers into compliance. In anticipation of new regulations that will apply to a far broader category of produce than the 2013 cantaloupe assignments, NASDA recommends a partnership between the FDA and the organizations with experience in on-farm inspections, the state departments of agriculture.

The NASDA referred to other food safety programs implemented by the FDA and the USDA as models for the FDA to consider when implementing the Produce Safety Rule. For instance, the NASDA pointed to effective food safety programs involving the states that were already in place for seafood, meat and poultry, and juice. As the NASDA wrote to the FDA: "These programs can provide models for moving forward, but the FDA must recognize that these programs also provide striking contrast to the current proposed regulation. The previous programs are related to controlled processing of products, rather than the growing & harvesting process contained in the produce safety regulation."

On September 16, 2014, the NASDA announced a new cooperative agreement with the FDA to provide critical information for planning and implementation of the FSMA Produce Safety Rule in partnership with state regulatory agencies.[23] The cooperative agreement will provide the funding and support necessary to determine the current foundation of state law, the resources needed by states to implement the Produce Safety Rule, as well as develop a timeline for successful implementation. "Our state partners have expertise in produce safety and unique knowledge of local food production activities, and thus have an essential role to play in helping to implement the FSMA produce safety rule," said Michael R. Taylor, the FDA's Deputy Commissioner for Foods and Veterinary Medicine.[24] The NASDA president, Vermont Secretary of Agriculture Chuck Ross, stated:

> The progress we have made in the past year towards a state-federal partnership with the FDA is incredible. This agreement further confirms the critical need to make sure the produce safety rule gets implemented correctly. NASDA will help the FDA develop and implement a national produce safety plan in a way that makes sense to the producers and processors that feed American consumers. NASDA remains fully committed to food safety and the successful implementation of FSMA.[25]

I would argue that a similar model is needed for food safety oversight of artisan cheesemaking. Individual states already conduct inspections and license cheesemakers working within their states, and have done so for a long time.

On February 21, 2014, Harvard Law School hosted a conference on food safety law. Denis Stearns, a law professor at the Seattle University School of Law, questioned whether the FSMA's strict regulation of farms was even needed. As he stated, "Lots of FSMA regulation has to do with what people do on the farm level, but not much has changed for the big processing operations doing the bagging and cleaning. One of the weaknesses of FSMA could perhaps be that it isn't focused on the right thing."[26]

Stearns's perspective transcends produce. Artisan cheesemakers would heartily agree.

Criminal Penalties

Have the FDA's new authority and powers under the FSMA gone too far? The answer would be a resounding yes, according to a white paper authored by the attorney Shawn Stevens. The white paper, titled "FDA's War on Pathogens," offers very interesting and cautionary perspectives.[27] Stevens writes that Congress ordered the FDA to overhaul food safety when it passed the Food Safety Modernization Act (FSMA). He warns that the FDA is now conducting microbiological profiling inside food processing facilities during routine inspections and testing vast amounts of food at retail. The FDA is also initiating criminal investigations against food companies, along with their executives, who distribute foods that have the potential to cause human illness. These investigations involve multiple cases where food company executives had no direct knowledge that their food products were causing illness or even had the potential to cause illness. Many of these criminal investigations involve *Listeria monocytogenes* found either in food processing environments or in food products in commerce. Previously, the FDA permitted food companies to occasionally detect *L. monocytogenes* within the processing environment, so long as the *L. monocytogenes* contamination was controlled and did not contaminate food contact surfaces or products. However, under the FDA's new approach through the FSMA, the failure to eliminate sporadic *L. monocytogenes* findings in the environment may now subject companies to criminal liability.

As Stevens writes:

> In the event any environmental sample tests positive, depending upon the location of the sample and the circumstances surrounding its collection, the agency may require the company to recall potentially affected product. In the event this occurs,

> recall exposure is not the only exposure facing the companies at issue. As FDA continues to perform extensive microbiological sampling in food production facilities, the agency will continue to perform genetic DNA testing on any positive samples collected from those facilities, and compare the DNA fingerprints of those samples against the DNA fingerprints of sick case patients over the last 20 years. If FDA discovers that the DNA fingerprint from any of those samples matches an illness (or illnesses) in the PulseNet database, FDA will immediately presume that all of the illnesses were caused by a food product that originated from that facility.

This chilling hypothetical became reality for Blue Bell Creameries, the Brenham, Texas–based ice cream facility. Using PFGE and whole genome sequencing, the CDC was able to match strains of *L. monocytogenes* from infected patients with ice cream products produced at Blue Bell production facilities.[28] Although the product contamination was identified in 2015, the patient cases that were linked to Blue Bell products spanned a time period of five years from 2010 to 2015.

Stevens writes that the FDA can also conduct microbiological sampling of foods collected from retail stores. For foods that test positive, in addition to mandating a recall of the affected product, FDA will also demand entrance into the production facility at issue and begin sampling the environment extensively (taking hundreds of samples from the drains, floors, walls, production equipment, and finished products) in an effort to find the same strain as the positive sample found at the retail location. The FDA has demonstrated its intent to initiate a criminal investigation against any food company executives or quality assurance managers involved in a case where a positive sample collected by the FDA from their food facility or product is linked to a foodborne illness.

Does threatening criminal penalties improve food safety? I know for a fact that the FDA's policies create fear. What is needed to help food producers in the production of safe foods is education and training. I know unequivocally that learning does not occur in a climate of fear. What fear does, however, is incentivize people to hide things from those providing oversight.

Essentially, the FSMA has now placed the focus of food safety within a legal, not a scientific, framework. Under the FSMA, the FDA now has mandatory access to production records at food processing establishments. What incentives has the FDA provided for companies to really seek out sources of *Listeria* in their production environments, knowing that positive findings will trigger the cascade of recalls, illness investigations, and criminal prosecutions? This system of negative rewards has actually made our food less safe. Fortunately, there is a better way forward.

CHAPTER 12

Timeline of Key Events and Hard Questions

As isolated incidents, the reexamination of the 60-day aging rule in 1996, the initiation of the Domestic and Imported Cheese and Cheese Products Compliance Program in 1998, the 2009 and 2010 Compliance Policy Guide (CPG) Sec 527.300 Microbial Contaminants and Alkaline Phosphatase Activity *E. coli* regulations, the wooden boards ban in 2014, the updated 2015 Domestic and Imported Cheese and Cheese Products Compliance Program *E. coli* criteria, and reexamination of a performance standard for raw milk cheese in 2015 created great concern and confusion, along with adverse economic impacts, for the United States as well as the global artisan cheese industry. Yet there was no scientific basis for these regulatory changes. It is notable that the 60-day aging rule remained unchallenged from 1950 onward until the trigger event in 1992, when the EU began intellectual property protection of its traditional cheeses through use of geographical indications (GIs). From that time forward, the FDA's regulatory activity toward cheesemakers mirrors the trade disputes that were occurring at the WTO, with the most contentious issues revolving around the EU's protection of common cheese names.

Despite comprehensive and transparent risk assessments conducted by FSANZ (Food Standards Australia New Zealand) in 2002 that proved that aged hard Swiss and Italian Grana cheeses, along with Roquefort, have an equivalent level of safety with cheeses made from pasteurized milk, and despite the scientific literature confirming that the 60-day rule was working to protect public health, did the FDA push forward on proposed rules and regulations that were chiefly intended to disrupt commerce? To what degree did the FDA ensure that unbiased scientific information was the basis for all regulatory decisions involving artisan, raw milk, and traditional cheeses? Why did the FDA initiate regulatory attempts to ban the use of raw milk in cheesemaking and ban the use of wooden shelves in cheese aging if these actions were not supported by the large body of science surrounding these issues?

If there were a scientific basis for these actions, they would not be disputed. And if the FDA's scientific justification of these issues had merit, the agency would have defended these policies; instead, it backed down when challenged by members of Congress. These regulations, both implemented and proposed, had a significant financial impact on cheeses exported from Europe and on US artisan cheeses, but these rules and regulations did little to protect public health. And did focus on these issues distract the FDA from the important public health issues upon which it should have been focused? Costly recalls of US dairy products and outbreaks of illness and deaths linked to *Listeria monocytogenes* contamination of dairy foods have impacted both artisan as well as industrial cheese producers, along with their colleagues in the allied dairy industry.

When the regulatory activity documented above is juxtaposed against the trade disputes that were simultaneously occurring, a disturbing picture is painted. The timeline shown in figure 12.1 displays the linkage between FDA's regulatory actions and concurrent trade issues, and raises some serious questions. Did the EU's protection of its traditional cheeses in 1992 lead to the FDA's reexamination of the 60-day aging rule and calls for mandatory pasteurization of milk used in cheesemaking? If the intended regulatory goal was to keep Appellation d'Origine Protégée (AOP) and Protected Designation of Origin (PDO) cheeses from the US marketplace, mandating pasteurization is a good way to accomplish this objective, as most traditional cheeses are made from raw milk. Was this strategy abandoned when FSANZ, through comprehensive risk assessments, proved that aged hard cheeses carry an equivalent level of safety to cheeses made from pasteurized milk?

In 2009, were proposed regulatory changes for *E. coli* criteria in cheese made in response to the European Court of Justice ruling that Parmigiano Reggiano's common name, Parmesan, could become a protected name over time? Or was establishment of stringent *E. coli* criteria another way to mandate pasteurization of milk used for cheesemaking because it is difficult for certain cheeses made from raw milk to meet the arbitrary <10 *E. coli* MPN/g standard?

Did the FDA use the Food Safety Modernization Act rulewriting process as a mechanism to target AOP and PDO cheeses? Why would a proposed ban on wooden boards in cheese aging suddenly appear in 2014? And why in 2015 did the FDA continue its focus on cheeses made from unpasteurized milk? It is evident that France was not the only country experiencing a David versus Goliath battle between its industrial and artisan cheesemakers. A similar battle was occurring here in the United States, but was this battle different? Were US industrial cheesemakers aided and abetted in their battle with the artisan cheese community both here and abroad by regulatory officials working within the US Food and Drug Administration?

Attendees at the American Cheese Society annual meetings began to notice some interesting connections between the dates when the annual ACS meetings were occurring and the timing of FDA's recall announcements and unannounced visits to their establishments. Cheesemakers would lament that they were unable to bring the quality assurance members of their organizations to the ACS annual meetings because it became a matter of routine that unannounced inspections from the FDA would occur during the ACS meeting. Other attendees would tell of cheese being taken by FDA compliance officers for microbiological compliance sampling, which meant that cheesemakers would need to withhold the production lots being tested from distribution into commerce until results of the FDA's analysis were returned. Affected companies noticed a pattern: Many times results were released only after the products had reached the end of their sell-by dates. And in most cases, the tested products met compliance criteria: They were salable products that fully complied with regulations, but they could not be sold, because of regulatory targeting and testing of these goods. The ACS reached out to the FDA on several occasions to establish a dialogue with the agency, but, often, the FDA would send representatives who had little expertise or knowledge regarding the issues in which ACS members had greatest interest. In 2015, FDA officials agreed to attend the ACS annual meeting held in Providence, Rhode Island, to address ACS members at a general session. As Dr. Susan Mayne of the FDA was speaking at the

TRADE ACTIVITY

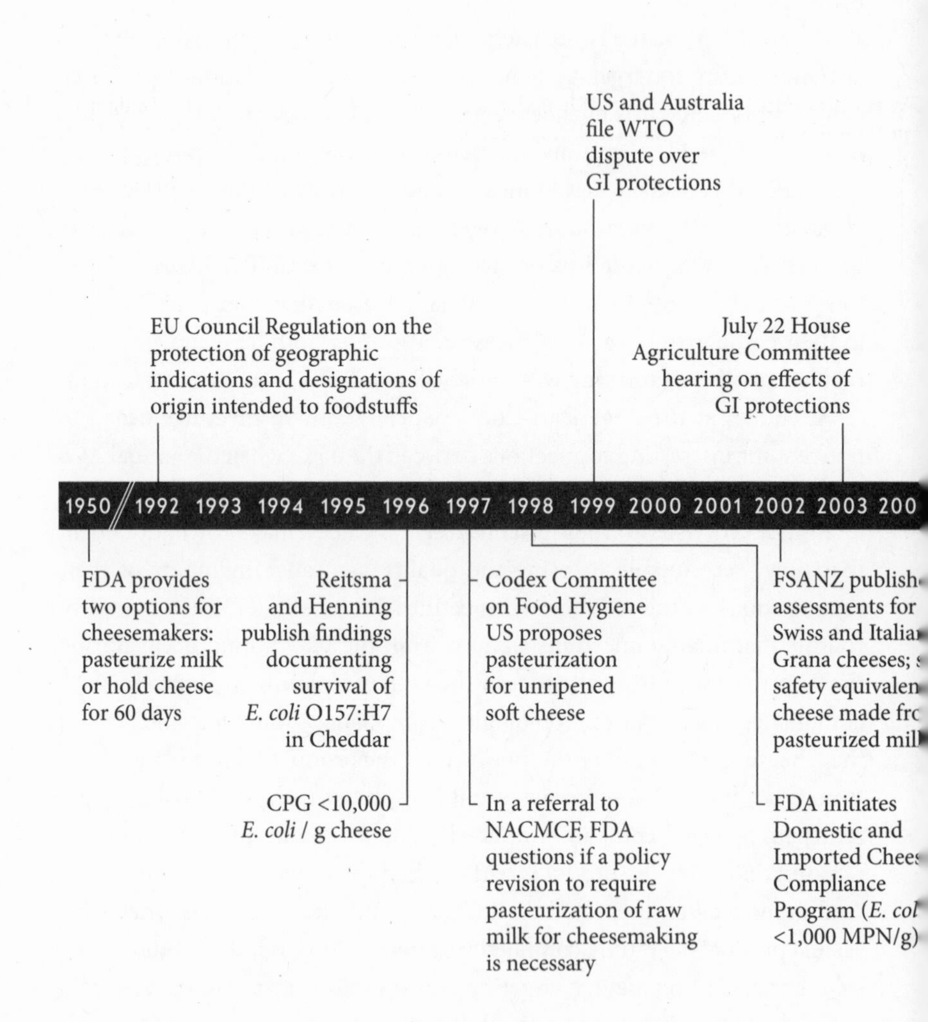

US and Australia file WTO dispute over GI protections
EU Council Regulation on the protection of geographic indications and designations of origin intended to foodstuffs
July 22 House Agriculture Committee hearing on effects of GI protections
1950 1992 1993 1994 1995 1996 1997 1998 1999 2000 2001 2002 2003 200
FDA provides two options for cheesemakers: pasteurize milk or hold cheese for 60 days
Reitsma and Henning publish findings documenting survival of E. coli O157:H7 in Cheddar
Codex Committee on Food Hygiene US proposes pasteurization for unripened soft cheese
FSANZ publish
assessments for
Swiss and Italia
Grana cheeses;
safety equivalen
cheese made fr
pasteurized mil
CPG <10,000 E. coli / g cheese
In a referral to NACMCF, FDA questions if a policy revision to require pasteurization of raw milk for cheesemaking is necessary
FDA initiates
Domestic and
Imported Chee
Compliance
Program (E. co
<1,000 MPN/g)

REGULATORY ACTIVITY

Figure 12.1. Timeline of events and linkage between the FDA's regulatory actions and concurrent trade issues.

EU Court of Justice rules in favor of Consortium of Parmigiano Reggiano that GI can become a generic name over time

WTO DSB rules in favor of US, Australia

EU updates quality schemes for agricultural products

EU-Mexico Trade Agreement protects GIs and restricts use of common food names

US and Australia claim EC did not fully implement DSB recommendations

Consortium for Common Food Names formed

New food safety agency proposed by Trump administration

005 2006 2007 2008 2009 2010 2011 2012 2013 2014 2015 2016 2017 2018

Quebec legalizes sales of raw milk cheese aged less than 60 days

FSMA becomes law

FDA pauses *E. coli* testing of cheese

FDA revises CPG: non-toxigenic *E. coli* cannot exceed 1,000 MPN/g in cheese made from raw milk

FDA proposes ban on wooden shelves in cheese aging

FDA publishes docket on potential health risks in cheese made from unpasteurized milk

FDA revises CPG; cheeses are adulterated if *E. coli* >100 MPN/g

Domestic and imported cheese and cheese products *E. coli* standards finalized

ACS meeting on July 31, 2015, and pledging cooperation of the FDA to work with ACS members, the FDA compliance division was busy at work releasing Compliance Program 7303.037, Domestic and Imported Cheese and Cheese Products, with an implementation date of July 30, 2015, which finalized the FDA's establishment of stringent *E. coli* criteria. This was done exactly 1 day before Dr. Mayne's remarks. Unfortunately, there was no mention of this food compliance program being finalized in Dr. Mayne's speech. Instead, Dr. Mayne announced the FDA's request for comments from the public that would assist the agency in identifying and evaluating measures that might minimize the impact of harmful bacteria in cheeses made from unpasteurized milk.

With respect to the FDA's FY 2014–2015 raw milk cheese sampling pilot assignment that began on January 1, 2014, results of this assignment were not released until July 21, 2016. Why did it take the FDA two and a half years to compile the results of its analysis of 1,606 raw milk cheeses? With all of the resources that the FDA has at its disposal, could the agency really only conduct an analysis of 800 cheese samples per year? In the final analysis, the FDA found that less than 1 percent of tested cheeses contained pathogens, and the FDA failed to detect *E. coli* 0157:H7 in any tested samples. But the FDA already had 16 years' worth of data from thousands of cheese samples tested since the initiation of the Domestic and Imported Cheese and Cheese Products Compliance Program in 1998. What were data from 1,606 samples going to provide over and above the voluminous information that the FDA had already collected? Of course, the revised and stringent *E. coli* standards remained in force during this two-and-a-half-year period, adversely impacting both domestic and imported artisan cheeses. Was the FDA strategy designed to delay findings and thereby cause prolonged economic harm?

On October 29, 2015, in response to the FDA's request for data "that would help the FDA in identification and evaluation of measures to minimize the impact of harmful bacteria in cheese made from unpasteurized milk," the ACS took the opportunity to respond to the request, but also used the opportunity to ask some hard questions of the FDA.[1] The ACS wrote, "ACS seeks to understand FDA's renewed focus on the production and sale of cheese made from unpasteurized milk." In response, the FDA justified this examination on public health grounds, stating its goal of "wanting to keep contaminated food from reaching consumers."[2] As stated by the FDA regarding its raw milk cheese sampling program, "The new approach, detailed in the Background section of this report . . . centers on the testing of a statistically determined number of samples of targeted foods over a relatively short period of time, 12 to 18 months, to ensure a statistically valid amount of data is available for decision making." Yet our timeline suggests that there were reasons beyond contaminated foods reaching consumers that explain why the FDA may have conducted this evaluation. There were a host of contentious trade issues ongoing during that time. The FDA wrote, "Of the 1,606 raw milk cheese samples collected and tested, 473 samples (29 percent) were domestic samples, and 1,133 samples (71 percent) were of international origin. The FDA sought to design its sampling plan to approximate the ratio of domestically made versus imported product on the U.S. market but was unable to do so in this case because the federal government does not track production volume of raw milk cheese."

The FDA further responded to the ACS:

> The FDA tested samples for the presence of the pathogens *Salmonella*, *Listeria monocytogenes*, *E. coli* O157:H7 and Shiga toxin-producing *E. coli*, as well as for generic *E. coli*. The overall contamination rate for each of the pathogens was less than one percent, and the overall contamination rate for generic *E. coli* was 5.4 percent. While the prevalence for generic *E. coli* was comparatively high, it bears mention that it rarely causes illness even as it may signal insanitary processing conditions.

Yet, between 2004 and 2006, the FDA had testing information on more than 17,000 cheese samples. If the number of samples tested for *E. coli* during these years was relatively representative of the number of cheese samples being tested by the FDA per year (approximately 1,000 per year), between the years 2004 and 2014, the FDA should have been able to access data from over 10,000 cheese samples.

It is notable that raw milk cheese was included in a compliance sampling that involved a range of produce commodities. According to the FDA, the raw milk cheese sampling assignment was part of a broader assignment,

"Testing to Support Prevention Under the FDA Food Safety Modernization Act."[3] The FDA indicated that the first year of this effort (2014) was focused on three commodities: sprouts, whole fresh avocados, and raw milk cheese (aged 60 days). Samples of each commodity were collected and tested for *Salmonella, Listeria monocytogenes,* and *E. coli* O157:H7. For FY 2016, 1,600 samples of cucumbers and hot peppers were included, with testing conducted for *Salmonella* and *E. coli* O157:H7. The FDA also tested hot peppers for Shiga toxin–producing *E. coli*. In FY 2018, the FDA sampled fresh herbs (basil, parsley, and cilantro) and conducted testing for *Salmonella*, Shiga toxin–producing *E. coli*, and *Cyclospora cayetanensis*. The FDA also sampled processed avocado and guacamole and tested these products for *Salmonella* and *L. monocytogenes*. According to the sampling assignment results, the only commodity tested for non-toxigenic *E. coli* was raw milk cheese. As the FDA indicated in its raw milk cheese summary report, "While the prevalence for generic *E. coli* was comparatively high, it bears mention that it rarely causes illness even as it may signal insanitary processing conditions." Why were avocados, sprouts, herbs, cucumbers, and hot peppers not tested for generic *E. coli* if it was such a meaningful indicator of insanitary processing conditions? The inclusion of raw milk cheese in this particular compliance sampling program raises yet more questions about the FDA's motivations.

The FDA's regulatory actions were costly not only for cheesemakers but also importers. The news website Food Republic reported: "In 2014, the FDA stopped French cheeses like Roquefort, St. Nectaire, Morbier, and Tomme de Savoie from coming into the country. Importer Shawn Hockert, sales manager at Seacrest Foods in Lynn, Massachusetts, says his company lost roughly $250,000 in sales in 2014 because of the ban on Morbier. In the meantime, his company has fought the ban on Roquefort and has tried to find a new product to replace Tomme de Savoie."[4] Cheesemaker Andy Hatch also questioned FDA's approach: "So the perplexing thing about what the FDA is doing is testing for and regulating bacteria that don't pose any health risk."[5] Yet this was not done for produce commodities. What was the justification?

The Specialist Cheesemakers Association (SCA), a British organization representing artisan cheesemakers, including some who had been adversely impacted by the FDA's *E. coli* standards, submitted a question to the FDA through the ACS. The SCA stated that FDA specifications require cheese not to exceed 10 MPN/g non-toxigenic *E. coli*. The SCA indicated that while this standard may be appropriate for pasteurized cheeses and raw milk hard cheeses with significant age, it presented a problem for soft and semi-soft cheeses made from raw milk. As the SCA stated:

> *E. coli* testing is used within the food industry to monitor gross contamination of products. However, the production of raw milk soft and semi-soft cheeses often involves long acidification times, during which levels of non-toxigenic *E. coli* may grow to detectable levels, even though the raw materials can be shown to contain <10 MPN/g *E. coli*. It is for this reason that there is no legal limit for non-toxigenic *E. coli* levels in raw milk cheese within the EU. While the Specialist Cheesemakers Association promotes a target of <1000 MPN/g in soft or slow-acidifying raw milk cheeses, we also recognise that in certain cases even higher levels are not necessarily indicative of a hygiene lapse. These cheeses are not made from compromised raw materials, nor do they present a risk to consumers' health. Why didn't FDA consider a more reasonable level of *E. coli* such as <1000 MPN/g in cheeses made from raw milk?

A response to the SCA's question can be provided from an analysis of data obtained by the Cheese of Choice Coalition's 2004 Freedom of Information Act request to the FDA for results from the Domestic and Imported Cheese Compliance Program. During our initial data analysis, my research group at UVM was interested only in pathogen incidence in cheeses, and the results of our analysis, published in 2011, was focused on reporting pathogen incidence in the tested cheese samples.[6] We returned to these data in 2014 to conduct a retrospective analysis of *E. coli* levels reported by the FDA from its analysis of cheeses sampled during the three-year period between 2004 and 2006. We presented the results of our analysis at the annual meeting of the International Association for Food Protection in 2016.[7] Out of 3,413 samples of cheeses tested by the FDA for non-toxigenic *E. coli* during the years 2004, 2005, and 2006, 2,584 samples (75.7 percent) contained *E. coli* levels that exceeded 10 MPN/g. Out of these samples, 2,287 (67.0 percent) exceeded the 2009 Compliance Policy Guide (CPG) criteria of 100 MPN/g. In comparison, out of 3,413 cheese samples tested, only 333 (9.8 percent) of tested cheese samples exceeded EU criteria of <1,000 *E.coli* / g for cheeses made from heat-treated milk. Of these samples, only 242 (7.1 percent) exceeded the FDA's 1998 Domestic and Imported Cheese and Cheese Products Compliance Program criteria of <10,000 MPN/g. Of the cheese types tested, Mexican-style soft cheeses, semi-soft cheeses, and soft-ripened cheeses were the cheese types most impacted by the 2015 non-toxigenic *E. coli* standards. Interestingly, there was no statistical correlation between cheese samples that tested positive for *L. monocytogenes* and non-toxigenic *E. coli* levels. Samples that tested positive for high levels of *L. monocytogenes* had associated non-toxigenic *E. coli* levels as low as <3 MPN/g. We concluded that application of the FDA's 2015 Domestic and Imported Cheese and Cheese Products Compliance Program non-toxigenic *E. coli* standards (<10 MPN/g) had the potential to significantly impact domestic and imported cheese commerce based upon an analysis of FY 2004–2006 program results, but with little impact on food safety. Since the FDA had these data from these three years and more, could this be the reason that such stringent *E. coli* standards were established? Was this retaliation for the European Court of Justice 2009 ruling that a GI can become a generic cheese name over time? The FDA stated in its Constituent Update on February 8, 2016: "Our surveillance sampling shows that the vast majority of domestic and imported raw milk cheeses are meeting the established criteria."[8] This statement may have been correct for the 1,606 raw milk cheeses that FDA analyzed between FY 2014 and 2016, after the CPG was changed in 2009 and again in 2010. But the FDA's own data from 2004 through 2006 confirmed that approximately 70 percent of tested cheese samples had *E. coli* levels that exceeded established criteria, and analysis of data from the years 2006 to 2009 would likely further confirm this fact. Why did the FDA choose to ignore the voluminous data in its possession and make statements that could not be supported by its own scientific results?

Finger Lakes Farmstead was the unfortunate recipient of an FDA warning letter issued in 2012 because FDA laboratory analysis of a finished product sample of Schuyler Gouda cheese "found high levels (sub #1 = 150 MPN/g, sub #2 = 93 MPN/g) of generic *Escherichia coli (E. coli)*."[9]

~

The International Dairy Foods Association (IDFA) is an organization that "represents the nation's dairy manufacturing and marketing industry." IDFA members range from multinational organizations to single-plant companies. Together they represent approximately 90 percent of the milk, cultured products, cheese, ice cream, and frozen desserts produced and marketed in the United States and sold throughout the world. Geographical indications (GIs) are an area of concern for IDFA members. As stated by the IDFA:

> The European Union wants to expand rules that identify products by where they originate—so-called Geographical Indications or GIs—as a way to increase its sales of cheese across the globe. This means that U.S. companies would no longer be able to sell cheese labeled with common names such as parmesan, feta and asiago in the U.S. and foreign markets. Currently, the EU is trying to win more market share by attempting to claim sole rights to use common names. The U.S. dairy industry does not oppose legitimate GIs, but the EU's push in trade negotiations for onerous restrictions in third-country markets goes too far. IDFA believes trade agreements should not impose restrictions like these on market access and intellectual property rights.[10]

John Sheehan, the director of FDA-CFSAN's Division of Dairy, Egg, and Meat Products and Monica Metz, chief for FDA-CFSAN's Milk and Milk Products Branch, were regular speakers at IDFA meetings. At IDFA's regulatory roundup, held on June 13 and 14, 2016, topics scheduled to be presented by Sheehan under the Food Safety theme were: upcoming FSMA deadlines; what inspectors will be looking for; results of past sampling assignments; notices of upcoming assignments; and changes in policies for dairy-related *Listeria monocytogenes*, and more.[11] It is curious that Sheehan never attended ACS meetings to share similar information, despite numerous attempts by the ACS to invite his involvement. The IDFA also scheduled presentations under the Business and Trade theme, with topics that included barriers to exporting to the European Union; current status of negotiations for the Transatlantic Trade and Investment Partnership (T-TIP); and how to export to China and other emerging markets. The close and frequent association of certain FDA regulators with organizations such as IDFA, and regulations that have followed that have created economic harm to artisan cheese producers, both domestically and internationally, are concerning. Since the IDFA represents the majority of industrial dairy producers in the United States, has the FDA provided industrial cheesemakers with a competitive advantage in the marketplace? And why would high-ranking FDA dairy regulatory officials such as Sheehan and Metz be unwilling to share the same regulatory information and updates with ACS members? Many ACS members are also members of IDFA. Food safety should be a noncompetitive issue amongst dairy producers, and FDA must ensure that regulations are fairly applied across the entire US dairy sector. To date, the FDA has failed in this regard.

Two FSMA framework rules are the Foreign Supplier Verification Program (FSVP) for Importers of Food for Humans and Animals and Accreditation of Third-Party Auditors/Certification Bodies to Conduct Food Safety Audits and to Issue Certifications. Under the FSVP regulations, importers are required to perform risk-based activities to verify that food imported into the United States has been produced in a manner that provides the same level of public health protection as that required for domestic food producers. But using what standards? Science-based criteria, or arbitrary FDA-imposed criteria? One of the FSMA mandates directs the FDA to establish a program for the Accreditation of Third-Party Auditors for foreign food facilities. Under the regulation, the FDA would recognize accreditation bodies based on certain criteria such as competency and impartiality. Then, the bodies would in turn accredit third-party auditors to conduct food safety audits and issue certifications for foreign facilities and food under specified programs.[12]

On March 13, 2018, Janet Fletcher wrote on her *Planet Cheese* blog about the unavailability of cheeses such as Abbaye de Belloc, Laguiole, St. Marcellin, Montbriac, Brie de Melun, and St. Nectaire *fermier*, cheeses that before FSMA had been widely available to US consumers.[13] The reason: the Foreign Supplier Verification Program and the paperwork involved, which became a burden for the monks at the Abbaye Notre-Dame de Belloc monastery. But beyond paperwork requirements, if the FDA is now imposing US microbiological standards on cheeses produced abroad and disregarding the microbiological standards in use in the EU that are consistent with ICMSF (International Commission on Microbiological Specifications for Foods) requirements, is this not an imposition of a trade barrier? Are we not then asking all foods produced abroad to conform with our US production requirements, and, in so doing, creating a singular global food system? If there isn't transparency here in the United States, how can FSMA possibly work on an international level?

In 2016 FDA Deputy Commissioner Michael R. Taylor wrote of FSMA:

> This new mandate for FDA must be seen in the context of similar efforts to strengthen food safety systems all over the world. In fact, one of the reasons this is such a historic time for food safety is the great degree of alignment that exists across the global food system on the need to build modern preventive measures into food production operations and, equally important, to verify that those measures are in place and working. The world's international food safety standard setting body—the *Codex Alimentarius* Commission of the United Nations—calls for that approach. Implementing that philosophy is what FSMA is all about.[14]

But as we have seen, we are not philosophically aligned on equivalent approaches to achieve food safety, and FDA cheese regulations are not harmonized with the *Codex Alimentarius* or ICMSF. If "modern preventive measures" in FSMA will be used to eliminate traditional cheesemaking practices that use raw milk in cheese production and ban the use of wooden tools and shelves, we will end up with a food system that disregards the contributions of local, healthy, and safe traditional products.

Taylor continues: "We have full-time staff working on international harmonization of regulatory standards, trade policy issues, and regulatory partnerships with foreign governments. Our international engagement—and the importance of international relations to our success—is only increasing." In my experience with the artisan cheese community, nothing could be further from the truth. What voice do small-scale food producers have in this important dialogue? Where and how are the interests of small-scale food producers represented at the table when such important matters are being discussed and negotiated globally? Taylor goes on to state:

> In addition to recognizing the global reach of large companies, Congress noted that prevalence of thousands of small-scale growers and embraced their value to the food system and the emergence in recent years of a vibrant local food movement. In some U.S. communities, this movement, which responds to consumer interest in buying food grown closer to home, provides central features of the social fabric and important contributions to economic sustainability and development.

I am certain that my colleagues in the artisan cheese community would find these comments to be interesting, and inconsistent with their FDA experiences. Taylor's comments offer the right spirit and intent, but as we have seen from the many examples offered in this book, actions taken by FDA's Milk and Milk Products Branch and the Office of Compliance in the Center for Food Safety and Applied Nutrition are threatening the very existence of artisan cheesemakers, both domestically and globally, who make enormous contributions to the vibrant local foods movement. The FDA's regulations have been promulgated despite scientific evidence showing that artisan practices can produce safe cheeses.

Speaking at the IDFA's Regulatory RoundUP conference in 2017, FDA's Deputy Commissioner for Food and Veterinary Medicine, Dr. Stephen Ostroff, cited the FDA's mantra, "Committed to educate before and while we regulate," as he explained that the FDA recognizes that implementing the new (FSMA) food safety regulations is a learning process that will take some time. He explained that 300 "educational inspections" would take place under the preventive controls rule, with inspectors visiting 240 domestic and 60 foreign supplier facilities. The inspections, he said, were meant to acknowledge what companies are doing well, as well as identify areas that companies could improve so as to come into compliance. But with the threat of consent decrees and criminal prosecutions, would artisan cheesemakers be helped or harmed by these visits? The FDA actions described in this book have created mistrust between artisan cheesemakers and the regulatory community. Mistrust does not serve to promote food safety. And the most unfortunate part of FDA's focus on artisan cheese is that artisan cheesemakers are unequivocally committed to the safety of the products they produce.

PART THREE

What Do We Do Now?

CHAPTER 13

Toward a New Regulatory Model

The FDA is not the sole agency in the United States charged with food safety regulatory authority. The FDA has regulatory jurisdiction for a wide variety of foods, including dairy products, produce, juice, and seafood. Meanwhile, the USDA Food Safety and Inspection Service (FSIS) has statutory regulatory authority for the safety of domestic and imported meat, poultry, processed eggs, and catfish. The USDA-FSIS's food safety responsibility includes everything from large plants processing millions of pounds of ground beef per day to very small meat manufacturers that make sopressata and other dry fermented, traditional specialty meat products. The FDA and USDA-FSIS have very different approaches and philosophies to food safety assurance. Perhaps the best example is the USDA-FSIS's approach to *Listeria* control to ensure the safety of meat products. USDA's approach differs substantially from that used by the FDA to ensure the safety of dairy foods, including cheeses made from raw milk.

Back in the late 1990s to early 2000s, a number of large multistate outbreaks of listeriosis were linked to contaminated ready-to-eat (RTE) meat products, including deli meats and hot dogs. For example, a 2002 outbreak of listeriosis linked to turkey deli meat sickened 54 patients, caused 8 deaths and led to fetal loss in 3 pregnant women.[1] The outbreak strain was found in the processing environment of one plant, and in turkey products and the processing environment of a second plant. More than 30 million pounds of products were recalled from the two implicated processing facilities, and this and other similar outbreaks led the USDA-FSIS to issue new regulations. To prevent future outbreaks, FSIS put in place a system that affords a much more flexible and much less punitive approach to *Listeria* control than that used by the FDA. And it works to control *Listeria* and thereby protect public health. A significant decline in outbreaks of illness and incidence of contamination of RTE meats has occurred as a direct result of collaborative efforts between the meat industry and academia, working through the North American Meat Institute, along with regulators working for the USDA-FSIS.

Like the FDA, the FSIS has maintained a "zero tolerance" policy for *Listeria monocytogenes* in RTE meat products, defined as products that are safe to consume without the need for further preparation, such as recooking. Hot dogs and deli meats are examples of RTE meat products to which the USDA-FSIS *Listeria* Rule applies.[2] RTE meat and poultry products are processed using a lethality step that includes either cooking or fermentation and drying to eliminate any pathogens that might be present in these products. After the lethality process has been applied, RTE products may become recontaminated with *L. monocytogenes*. USDA Regulation, 9 CFR 430.4(a) states that *L. monocytogenes* is a hazard that must be controlled by companies producing RTE products that will be exposed to the post-lethality environment.[3] This is done through a Hazard Analysis and Critical Control Points (HACCP) plan that identifies points in processing where hazards such as *Listeria* might be introduced, and through sanitation and other programs. RTE products are considered adulterated if they either contain *L. monocytogenes* or if they come into direct contact with a food contact surface that is contaminated with *L. monocytogenes*.

The regulation provides processors with the flexibility of choosing one of three alternatives to meet the regulatory requirements. Under Alternative 1, if a processing establishment applies both a post-lethality treatment to reduce or eliminate *L. monocytogenes*, along with an antimicrobial agent or process (AMAP) to control *Listeria* growth, this product poses less of a risk for *Listeria* contamination. In turn, the USDA will subject the company to less stringent compliance sampling than would be done if neither of these treatments was used. Following slicing and packaging of processed meats, examples of post-lethality processes include applying steam or hot water pasteurization to package surfaces or subjecting packaged cold cuts or hot dogs to high hydrostatic pressure processing, a nonthermal process that destroys bacteria. Processors can also elect to reformulate their products to inhibit *Listeria* growth. Research championed by the Oscar Mayer division of Kraft Foods found that reformulating hot dogs or deli meats with sodium lactate or potassium diacetate could inhibit growth of *Listeria.*[4] Under Alternative 2, processors apply either a post-lethality treatment or an antimicrobial agent. They would receive more stringent compliance sampling than a processing facility using Alternative 1. Using Alternative 3, the establishment does not apply any post-lethality treatment or antimicrobial agent or process and instead it relies on its sanitation program to control *L. moncytogenes.* These plants in turn would be sampled more stringently than those using Alternatives 1 and 2. This is an example of risk-based compliance that provides processors of all sizes with flexibility for controlling a significant potential hazard presented to their products.

USDA-FSIS's science-based approach controls *Listeria* and other pathogens in a fermented meat product such as a dry fermented sausage. Pathogens such as *Listeria*, *Salmonella*, and Staph are much more likely to be present in raw meat used to make dry fermented sausage compared with raw milk used to make cheese. If processes such as fermentation and dry aging can allow the safe production of a fermented sausage, the FDA's approach to raw milk cheese safety seems nonsensical. FSIS guidance acknowledges the "hurdle concept," whereby salt, nitrites, and other additives achieve a water activity, pH, or moisture-protein ratio that will reduce the level of *L. monocytogenes* and other pathogens during processing, and will continue to inhibit the growth of the pathogens during the refrigerated shelf life of these products. The added salts and nitrites work together to create hurdles to pathogen growth. As applied to cheesemaking, if raw meat can be used to make microbiologically safe sausage through fermentation and loss of moisture, surely this can be done using raw milk in cheesemaking, where similar hurdles create microbiologically safe products.

The USDA-FSIS regulatory model offers a prime example of public health achievements that can be advanced in other food products. A significant decline in outbreaks of illness and incidence of contamination of RTE meat products occurred as a direct result of coordinated efforts by the food industry and regulatory agencies working in collaboration. Such efforts included improved manufacturing practices that minimized the risk of cross-contamination. Additionally, products were formulated with compounds (for example, sodium lactate / potassium diacetate) that limited or prevented *L. monocytogenes* growth. The incidence rate for *L. monocytogenes* in USDA-tested RTE products decreased as a result of changes in regulatory policy.

The USDA's view on the presence of indicator organisms, such as *E. coli*, in products also differs substantially from that of the FDA. As stated by the USDA:

> While not defined in 9 CFR 430.1, the term indicator organism is used in 9 CFR 430. Indicator organisms are bacteria used to determine objectionable microbial conditions of food, such as the presence of potential pathogens, as well as the sanitary conditions of food processing, production areas, or storage rooms. Lm belongs to the genus *Listeria* and the species is *monocytogenes*. The genus *Listeria* includes other nonpathogenic species (spp.) in addition to the pathogenic species *monocytogenes*. A positive test for *Listeria* spp. on a food contact surface would indicate the potential presence of Lm. However, the product is only considered adulterated if Lm is found on a food contact surface or product. If *Listeria* spp. is found, the product is not considered adulterated, however the establishment is expected to take corrective action, according to their control alternative, to address *Listeria* spp. positives so that the product does not become adulterated. If a test is negative for Lm or *Listeria* spp., this indicates Lm is not present. Note that tests for other indicator organisms, like aerobic plate counts (APC), total plate counts (TPC), and total coliforms are not appropriate indicators for Lm. Although such tests could provide a measure of general sanitation, they do not indicate the potential presence or absence of the pathogen of concern.[5]

This example illustrates how differently our two primary federal regulatory agencies approach safety assurance of the products over which they have regulatory authority. Coliforms referred to in this guidance are a group of bacteria that include *E. coli*. The FDA has defined low levels of *E. coli* as adulterants in cheeses, yet unlike the USDA, has not defined the science behind its rationale.

Further, the FDA is using the presence of indicator organisms to issue injunctions against artisan cheese manufacturers. For instance, on August 8, 2014, the US Department of Justice (DOJ) filed suit against S Serra Cheese Company, a Michigan-based artisan cheese firm, and its owners to prevent the distribution of allegedly adulterated Italian artisan cheese products.[6] The Department of Justice alleged that these cheeses were manufactured under insanitary conditions, as evidenced by the identification of two potentially dangerous types of bacteria during 2013 inspections of S Serra's manufacturing facility, and that the company's procedures were inadequate to ensure product safety. The inspections allegedly uncovered strains of *E. coli* and *Listeria innocua* that, while nonpathogenic, indicated that the facility, according to FDA, was unsanitary, contaminated, and could support the growth of life-threatening bacteria.

As stated in a suit filed by the US Department of Justice: "According to the complaint, two FDA inspections performed in 2013 revealed that the company's cheese is adulterated within the meaning of the Food, Drug and Cosmetic Act because it is prepared, packed or held under insanitary conditions in which it may have become contaminated with filth or rendered injurious to health." The complaint alleges, for example, that the company repeatedly failed to reduce the risk of contamination from these two potentially dangerous types of bacteria (*E. coli* and *Listeria innocua*).

As stated by DOJ's Office of Public Affairs in its press release issued on August 8, 2014: "Although the strains of *E. coli* found in cheese samples collected from the company's facility were non-pathogenic, their presence indicates that the facility is insanitary and contaminated with filth. In addition, the presence of *L. innocua* indicates insanitary conditions and a work environment that could support the growth of *L. monocytogenes*, an organism that poses a life-threatening health hazard because it is the causal agent for the disease listeriosis, a serious encephalitic disease. The presence of *L. innocua* in the company's facility demonstrates the potential for the presence of *L. monocytogenes* in the same processing environment." In the press release, Department of Justice Assistant Attorney General Stuart Delery stated, "The presence of potentially harmful pathogens in food and processing facilities poses a serious risk to the public health. . . .

The Department of Justice will continue to bring enforcement actions against food manufacturers who do not follow the necessary procedures to comply with food safety laws."[7] If the presence of indicator organisms is now the microbiological standard under FSMA that the FDA will apply to all FDA-regulated foods in commerce, the ripple effects across the entire food industry will be significantly felt, in the absence of any scientific information that our food will be safer. As we have already seen, these are positions that are not scientifically supported by either USDA-FSIS or the FDA's own NACMCF (National Advisory Committee on Microbiological Criteria for Foods).

An even more egregious example of overreach comes from the March 20, 2014, FDA consent decree issued to Finger Lakes Farmstead. The FDA, working with the Department of Justice, filed a complaint for injunction in federal district court on January 22, 2014, against Finger Lakes Farmstead and owner Nancy Taber Richards. On page 12 of the consent decree, the Department of Justice states:

> If any laboratory test completed pursuant to paragraphs 6 (c) (1)-(4) shows the presence of pathogens, including L. mono, or non-pathogenic E. coli at levels greater than 10 most probable number (MPN) per gram in two or more subsamples, or greater than 100 MPN per gram in one of more subsamples, in any article of food, the Defendants must immediately cease production and notify FDA that production has ceased. Defendants shall also destroy, at Defendants' expense, under FDA supervision, and according to a destruction plan submitted in writing by Defendants and approved prior to implementation, in writing, by FDA, all food products manufactured from the time the laboratory sample(s) testing positive for pathogens including L. mono, or non-pathogenic E. coli at levels greater than 10 most probable number (MPN) per gram in two or more subsamples, or greater than 100 MPN per gram in one of more subsamples.[8]

Under USDA regulations, neither *L. innocua* nor nonpathogenic *E. coli* are considered to be adulterants. Yet, for FDA-regulated foods, injunctions are being filed in federal court using the presence of indicator organisms as evidence of food safety violations. Is this food safety or government overreach? Why are federal rules and regulations so inconsistent, and what can be done about this? In response to an import alert issued by FDA to producers of Roquefort, Morbier, and other French cheeses, Janet Fletcher wrote about the impact that the FDA's arbitrary non-toxigenic *E. coli* standards were having on French cheese imports. She noted that cheeses on the import alert cannot be legally sold in the United States until the producers have documented that they have taken corrective action, and this process may take several months. As Ben Chapman wrote on *barfblog*, upon his review of FDA's 2009 Compliance Policy Guide (CPG): "Any government agency needs to clearly and effectively communicate risk-based decisions (especially changes) and provide the evidence to back a particular decision."[9]

In August 2015, the Alliance for Listeriosis Prevention sent a letter to the FDA urging the agency to consider examining existing USDA-FSIS industry guidance and for the FDA to consider adopting a more harmonized regulatory policy with that of the USDA-FSIS.[10] Specifically the Alliance was advocating for an FDA regulatory environment that allowed and encouraged environmental monitoring for *L. monocytogenes* indicators (such as *L. innocua*) to facilitate the use of science-based preventive control strategies to control *L. monocytogenes* in ready-to-eat foods. Additionally, the Alliance provided comments to the USDA-FSIS expressing support for the USDA-FSIS approach as established in the final guidance to industry for *L. monocytogenes* in post-lethality-exposed ready-to-eat meat products. The Produce Marketing Association, in a November 2015 letter to the FDA also supported the USDA's approach, writing:

> A regulatory environment that encourages aggressive environmental monitoring for *Listeria* indicators is what is required to facilitate the use of science-based preventive control strategies to control Lm in ready-to-eat foods such as fresh produce. As such we at PMA believe and are advocating for sound public policy that encourages food facilities to be able to proactively seek out and correct potential Lm harborages on food contact surfaces and non-product contact surfaces. One such regulatory model that FDA should consider is the United States

> Department of Agriculture (USDA) Food Safety Inspection Service (FSIS) "Compliance Guidelines to control *Listeria monocytogenes* in post-lethality exposed ready-to-eat meat and poultry products." The aforementioned DRAFT USDA FSIS policy guidance provides the meat industry with regulatory flexibility that encourages the use of the "seek and destroy" strategy when transient positive detections of *Listeria* species or *Listeria*-like organisms occur. Additionally, it would be very beneficial to the food industry if FDA and USDA FSIS had consistent approaches to environmental testing for *Listeria* especially in dual jurisdiction food facilities.[11]

The raw milk cheese debate should give us pause as we consider the FDA and the enormous power over the US food supply this agency has been granted under the Food Safety Modernization Act. Already, we have seen abuses to that power. If artisan cheese offers an example of how commodities will be regulated in the future, how will produce be impacted? Will arbitrary non-toxigenic *E. coli* standards be established for produce? Will use of wood in produce packing crates be arbitrarily banned? The FDA makes critically important decisions every day about drugs, medical devices, and veterinary safety. Is the behavior we have seen in the FDA's Milk and Milk Products Branch and the Office of Compliance in the Center for Food Safety and Applied Nutrition the new scientific standard this agency is upholding? It is chilling to think about. One expects tension between industrial giants and small-scale food producers who reflect cultural differences, difference in practices, and difference in philosophy and approach. However, we must expect integrity from those who regulate our foods and we must ensure that our regulators are transparent and that they are using the best and most credible scientific data in supporting their regulatory decisions. The raw milk cheese issues highlighted in this book should concern each and every American consumer. If FDA staff cannot correctly interpret the scientific literature, or provide members of Congress and their constituents with relevant, meaningful scientific information; or if they are allowed to make changes to regulatory standards to cause economic harm and disruption to food producers without consideration of internationally adopted *Codex Alimentarius* principles and ICMSF food testing standards, our food future here in the United States is in great jeopardy indeed. The FDA had to back down from *E. coli* regulatory standards applied to raw milk cheese, its position on wooden boards, and its position on cheese aging because the agency had no credible scientific support for its proposed or implemented regulations. These issues should have been resolved before establishing these regulations in the first place. If these policies went through layers of rigorous review at the FDA, clearly, no one in the agency had the expertise to realize that these positions could not be scientifically justified or defended. If we do not have transparency here in the United States, how can we possibly expect the FSMA Foreign Supplier Verification Program and the FDA inspection of food processing facilities abroad to serve food safety and consumer interests here at home?

To that end, we need a new system of food regulation in the United States that brings together knowledgeable scientific professionals to solve problems. In March 2017, President Trump tasked the Office of Management and Budget (OMB) to produce a comprehensive plan for reforming and reorganizing the government "to ensure that our organizational constructs are well aligned to meet the needs of the 21st Century." On June 22, 2018, the report "Delivering Government Solutions in the 21st Century" was issued. In this report was a proposal that deserves serious consideration and, in my view, support:

> To provide better food safety for the country and improve efficiency for stakeholders, the Administration proposes to consolidate core Federal food safety responsibilities into a single agency under USDA, where food safety is a top priority from farm to fork. This consolidation will give USDA the clear mandate, dedicated budget, and full responsibility it needs for optimal oversight of the entire U.S. food supply. Resources at the FDA will be freed up to focus on its core responsibilities of drugs, devices, biologics, and tobacco. Most importantly, this proposal will provide better food safety outcomes for the American people over the long term.[12]

As stated in the report:

> USDA is well poised to house the Federal Food Safety Agency. USDA is a strong leader in food safety; has a thorough understanding of food safety risks and issues all along the farm to fork continuum; and many agencies within USDA focus on food safety. The Agricultural Research Service (ARS) spends about $112 million on in-house food safety research, and ARS scientists work with both FSIS and FDA to help develop research priorities and food safety practices. In addition, many other programs at USDA have food safety elements, from helping to manage wildlife on farms, to monitoring animal health, to collecting pesticide residue data on fruits and vegetables. USDA also has established relationships between State departments of agriculture, local farms, and processing facilities, and is thus keenly aware of food safety issues at all levels.[13]

By way of background, in January 2015, Senator Dick Durbin and Representative Rosa DeLauro introduced the Safe Food Act of 2015. This act was intended to establish a Food Safety Administration (FSA) as an agency independent from the FDA. The FSA would administer and enforce food safety laws, with authority for all food safety oversight and associated activities including inspection, enforcement, and labeling.[14] President Obama's fiscal year 2016 budget, released on February 2, 2015, included a similar proposal for establishment of a Food Safety Agency. While the agency would be separate from the FDA, it would be housed within the Department of Health and Human Services (HHS), as was suggested in the Safe Food Act of 2015. But as we have seen in the numerous examples set out in this book, housing this administration under HHS is fraught with problems, and the regulatory overreach would likely continue, without there being a focus on food safety. A proponent of President Obama's proposal was Secretary of Agriculture Tom Vilsack. During an appropriations hearing, he defended the proposal as a way to streamline food inspection and ensure the safety of American consumers: "It's not about tradition. It's not about turf. It's about food safety. We have a system that no one can contend is as effective or efficient as it needs to be."[15] Might this inefficient and ineffective system also be subject to misuse by regulatory officials who wish to use food safety as the guise for constructing artificial barriers to free trade?

There are compelling arguments to support placement of the federal Food Safety Agency within USDA. The USDA regulatory approach supports companies' staying in business, as opposed to shutting them down. Will we be less likely to see entire industries impacted, such as occurred with the US tomato industry in 2008 and the romaine lettuce industry in 2018, under USDA authority? I certainly believe so. Since the Rural Business Development Grant Program, the Rural Business Enterprise Grant Program, the Agricultural Marketing Service and associated National Organic Program, the Economic Research Service, the Farm Service Agency, and the Foreign Agricultural Service are at the heart of USDA's mission, small food artisans here and abroad are more likely to receive support from USDA when placed within food safety oversight from FSIS. Since 2005, USDA Rural Business Development programs have invested $8.8 million in Vermont's artisan cheese sector. Vibrant food systems need vibrant rural economies. There is more of a chance to have a vibrant and diversified food system when food safety is placed under USDA's watch, in my opinion. And the sooner we facilitate this reorganization, the better.

CHAPTER 14

International Perspectives

Vibrant cheese cultures exist in many parts of the world. I hope we can continue to build one here in the United States. The best way for the United States to address our dairy trade imbalance with the EU is not through a trade war, but through development of outstanding cheeses here in America that consumers want and for which they are willing to pay. American tastes are rapidly changing, and we are fast becoming a nation of obsessed foodies. Unlike people of my generation, who were content to travel to Europe and, on their return, purchase a food with characteristics similar to what they had consumed abroad, today's obsessed foodies want to know if a cheese is PDO or AOP, or, if not, if it is made locally with a closed dairy herd; if the animals are grazed on pasture and if they are treated humanely; and if the producer is using sustainable practices that minimize food waste and reduce the carbon footprint. A return to the small farm culture will accomplish many of these ideals and more, and in the process provide us access to some great cheeses.

There is great consumer demand for artisan cheese, and it is an economic driver for rural communities at a time when depressed milk prices are forcing many dairy farmers out of business. State departments of agriculture provide oversight of cheese production and license producers, ensuring safe production that could be improved only through coordinated research, education, and training. Sadly, until recently, we have never enjoyed a bona fide cheese culture in the United States. We invest little to no money in research on artisan and traditional cheese varieties, unlike France and Switzerland, where millions of dollars are invested. I think there is a real opportunity here that we are missing as a country in returning to the small farm culture and locally produced, high-quality cheeses that consumers crave. While international companies including Emmi and Lactalis have purchased US artisan cheese companies to further enhance their market share, putting all of our US artisan cheese development in the hands of multinational companies is not advisable. Even French raw milk cheesemakers are crying foul. As Véronique Richez-Lerouge, founder of France's Unpasteurised Cheese Association, states: "The big industrial producers will not tolerate the existence of other modes of production. They are determined to impose a bland homogeneity upon the consumer—cheese shaped objects with a mediocre taste and of poor quality because the pasteurisation process kills the product."[1]

I think there is a better way forward. Let's allow these enterprises to continue to grow and flourish as family farms within the rural communities that will benefit from their presence. Let's level the playing field, spread the wealth and divide it equitably among all producers of artisan cheese, and not limit economic opportunities to a chosen few large industrial players. Jeanne Carpenter of the blog *Cheese Underground* writes: "According to the Census Bureau's 2012 Economic Census, between 2007 and 2012, the total number of cheesemaking establishments in the U.S. rose by 13 percent to 542, while growth in small establishments (defined as employing up to 19 people), rose more than double that rate, by 28 percent, to 250."[2] Dick Groves of the publication *Cheese Reporter* states: "Statistics are all well and good, but there's nothing like spending four days over the course of two summers examining the appearance, texture, aroma and flavor of an incredible array of high-quality cheeses to remind me that the specialty/artisan/farmstead cheese industry is alive and well, and has an extremely bright future ahead of it."[3] As a longtime participant as a cheese judge at ACS myself, I could not agree more with this assessment. But the artisan/farmstead industry has a bright future only if it is not subject to arbitrary regulatory rules and regulations that threaten its very existence.

Traditional US dairy cooperatives are beginning to see the many opportunities in the transition to artisan cheese development and can offer some novel models to support small-scale businesses. Cabot Creamery in my home state of Vermont began development of a traditional clothbound Cheddar cheese back in 2003. The cheese was the brainchild of May Leach,

Cabot's longtime director of quality assurance. With milk sourced from a single dairy herd owned by George Kempton in beautiful Peacham, Vermont, the milk is transformed into green (unripened) cheese in Cabot, Vermont, then transported to the Cellars at Jasper Hill, where it is coated with lard and wrapped in cloth. This cheese made its debut at the American Cheese Society Annual Conference in 2006, and won first place overall best in show honors—the top prize awarded by the American Cheese Society.

Cabot Clothbound Cheddar has inspired the production of many similar products by other artisan cheese producers located throughout the United States. This cheese is the mainstay of the Cellars at Jasper Hill and remains a proud collaboration between a traditional dairy cooperative and an enterprising farm devoted to artisan cheesemaking and affinage. Cabot Creamery proudly celebrated its 100th year in business in 2019. Despite its historical roots and traditions, Cabot is a progressive cooperative that collaborates, rather than competes, with artisan producers. The result is a win-win for the company, for its collaborators, for the rural communities in which these businesses are located, and for consumers who get to enjoy these incredible products. There is only one downside: There is more demand for Cabot Clothbound Cheddar than the company can deliver. On the Cabot website, the company has posted most frequently asked questions (FAQ). One of the FAQs is: "Why is it so hard to find the Cabot Clothbound cheese?" The response: "The unique aging process of our Clothbound Cheddar cheese limits the quantity available to our customers."[4] This is such a good problem for a dairy cooperative to have.

The Grafton Village Cheese Company located in Grafton, Vermont, also enjoys a long history from its founding as a dairy cooperative in 1892. The Windham Foundation now owns Grafton Village Cheese as part of its mission to promote Vermont's rural communities. The company is famous for its traditional aged Cheddar cheese made from the milk of Jersey cows sourced from local dairy farms. Recently, it added Shepsog, a natural-rinded cheese made from raw sheep and cow's milk that is aged for four to six months. Jasper Hill, Cabot, and Grafton Village Cheese are but 3 of the 49 member companies of the Vermont Cheese Council, who collectively produce over 150 different types of cheese on farms in Vermont, thereby promoting the working landscape and the rural beauty enjoyed by our state. Each year these products are showcased at the Vermont Cheesemakers

Festival at Shelburne Farms, a celebration of Vermont cheeses attended by over 2,000 cheese enthusiasts. Vermont artisan cheeses are revitalizing rural economies and creating jobs and economic development. Tourism spurred by the Vermont Cheese Trail and the companion world-class craft beers, ciders, wines, and distilled spirits that dot the Vermont landscape are creating a model foodie heaven. Specialty cheese remains the fastest-growing segment in the dairy industry. As my friend Mateo Kehler observes, "Over the next 20 years, the trend in the mix of domestic and imported cheeses on counters and in deli cases across the country is likely to flip-flop, and the potential for growth in this space will be healthy for many years to come." That is, unless the FDA has its way, threatening the market potential for $100 million worth of artisan cheese in Vermont over the coming years, produced by a diverse group of new and existing, large and small cheesemakers. What if instead this potential is handed over to multinational companies without the loyalty to the great state of Vermont or other states such as California, Washington, and Wisconsin, where such acquisitions are occurring?

As the artisan cheese industry continues to grow, we must make sure that these products are safely produced. We can turn to examples outside of our borders where cheesemakers in countries outside of the United States have developed an array of approaches to ensure cheese safety. Some of these examples are worth consideration as we move forward with new and improved regulatory systems here in the United States. In his November 30, 2018, column in the *Cheese Reporter*, my colleague Dan Strongin wrote a beautiful piece about his recent experience attending the Brazilian Artisan Cheese Awards.[5] Participating as a coordinator of Brazilian judging the 12 hand-selected and trained judging teams, the competition used the American Cheese Society's judging format that consists of pairing a technical judge with a sensory judge during evaluation of the 481 artisan cheeses entered in competition. The job of the sensory judge is to add points for positive cheese attributes, while the technical judge looks for defects and subtracts points. Being a coordinator of judging afforded Strongin the opportunity to reflect on changes in the structure of the Brazilian cheese industry. In contrast to the United States, which lost many of its artisan producers when cheesemaking became industrialized, Brazil maintained its small-scale cheesemaking traditions through a large network of small family farms that make cheese. Brazil had followed the lead of the United States in 1950 and mandated that all cheese made from unpasteurized milk had to be aged for 60 days or longer. This requirement devastated family farming. As Strongin describes:

> Many of these families went underground. Brazil is an underdeveloped country. The level of rural poverty is high . . . the law dealt a cruel blow to the many thousands of small farmers who were forced to go underground to survive and remain on their farms. Ironically, the dominant social challenge Brazil has faced for decades has been the migration of people from the countryside to the cities, forced to live in increasingly inhumane conditions within and around the major cities.

Fortunately, the 60-day aging law was challenged in the 1980s by Brazil's artisan cheesemakers. Strongin cites international agreements brokered by the WTO as the vehicle by which countries such as Brazil and others around the globe are moving toward more effective and less centralized sanitation controls, ones that are based on good practices and continual improvement and are much more effective than top-down bureaucratic policies, such as those used by the FDA, as barriers to trade. From a global view as mediated by WTO, Strongin notes that member states involved in international trade are obligated to provide the scientific basis for any food regulation, and to move away from the current regulatory model that promotes antagonism and dependence on inspection "to collaboration between authorities and producers and defining and managing the local risks during cheesemaking." Strongin comments: "Today, even the FDA has called into question the effectiveness of using time of maturation to ensure safety, and has suspended non-pathogenic coliform counts as a measure of the relative hygiene in the milking parlor or cheese room for artisan cheese." But, as we have seen in this book, the FDA would not have backed away from its stance on non-toxigenic *E. coli* without the advocacy of US artisan producers, scientists, and members of Congress who questioned the scientific validity of FDA's use of non-toxigenic *E. coli* standards as the basis for determining that cheese products were adulterated. Brazil's new federal law, the law of the Selo Arte, (the Artisan Seal), requires the simplification of regulatory hurdles used for artisan producers to pass inspections and the unification of rules and approvals into a single system that is regulated by the states. "For the first time producers currently registered and inspected by the states have the right to sell their traditional artisan products of animal origin throughout Brazil; though the implementation is lagging, it is moving forward." Proof that this system works is exemplified when 481 artisan cheeses from all over Brazil can enter a national cheese competition. Strongin notes that the gastronomic journalist and cheese judge at this competition, Eduardo Tristão Girão, wrote: "Yesterday, I and the other judges of the Brazilian Artisan Cheese Awards were deeply moved. In the final stage of the contest, after looking at excellent cheeses from all over Brazil, we had to select the winner in the Super Gold category, the best of the best. Moreover, when we decided that the winner would be a creamy Búffalo Milk Cheese from Saint Victor Farm, in the municipality of Salvaterra, on the Island of Marajó (PA), many cried."[6]

In the UK, the Specialist Cheesemakers Association (SCA) is an alliance of organizations involved with artisan cheese that includes cheesemakers, retailers, wholesalers, and others. The SCA was established to encourage excellence in cheesemaking. The association provides a forum for its members to exchange ideas, and it represents member interests to the British government and media. The SCA helps retailers and cheesemakers alike navigate technical, health, or hygiene issues. Members must demonstrate their commitment to quality and safety through adherence to the SCA Assured Code of Best Practice. Established in 1989, the SCA is most fortunate to have as its patron the Prince of Wales, HRH Prince Charles. As in the United States, regulatory authorities in the UK have challenged the practice of using raw milk in cheesemaking.

Regulation (EC) 178/2002 establishes the general principles of food safety and food law, aimed at preventing the marketing of unsafe food and ensuring that systems exist to identify and respond to food safety problems.[7] Article 5 of Regulation (EC) No 852/2004 requires food business operators (FBOs) to put in place, implement, and maintain permanent procedures based on Hazard Analysis and Critical Control Points (HACCP) principles.[8] As we have seen, HACCP procedures are internationally recognized as useful tools for FBOs to control hazards that may occur in food (EC Commission Notice 2016/C 278/01).[9] Regulation (EC) No 853/2004 sets criteria for raw milk production and cheesemaking. In 2015 the SCA's Assured Code of Practice (ACoP) recommended more stringent microbiological criteria, including *S. aureus* counts of <100 cfu/ml for raw cow's milk with absence of both *Salmonella* and *Listeria monocytogenes* per 25 ml of tested raw milk. The SCA Assured Code of Practice discusses the legal requirement for HACCP in the UK and the need to control hazards through identification of Critical Control Points (CCPs). CCPs must be monitored each time cheese is made to ensure that the cheesemaking process is controlling identified hazards, or whether corrective actions are needed to bring the cheesemaking process back under control. The SCA indicates the need for an optimum acidity profile for each unique cheese type produced, along with a description of corrective action that will happen if deviations from this profile occur.[10] Initial cheese composition targets (pH, Aw, salt-in-moisture) should be measured on a specified date, for example, 7 days post-manufacture. The SCA describes how sampling should be accomplished, ensuring that samples are representative.

Educated cheesemakers are safe cheesemakers. Cheesemakers in Europe, Switzerland, and France produce iconic cheese varieties including Gruyère, Comté, and Vacherin Mont d'Or within regulatory guidelines that simultaneously address food safety concerns while protecting traditional production and affinage practices. We discussed in earlier chapters of this book Australia's challenge to the government of Switzerland to document the safety of its cheeses exported to Australia. Raw milk cheese production in Switzerland comprises 80 percent of total cheese production, of which 30 percent is exported to the EU. In 2003, the Swiss federal government established the National Monitoring Program of Milk and Dairy Products (NMPD) in response to impending EU trade regulations that would impact Swiss cheese exports.[11] The purpose of the monitoring program was to ensure that all Swiss cheese exported was in compliance with EU regulations from 2003 to 2011, before monitoring was incorporated into a national plan. The program assessed the prevalence of microbial hazards in dairy products and identified which products and practices pose a higher risk of contamination above a microbiological threshold by targeting *Salmonella* spp., *Campylobacter* spp., Shiga toxin–producing *E. coli*, *L. monocytogenes* (LM), and the family of coagulase-positive staphylococci (CPS) encompassing *Staphylococcus aureus*. NMPD sampling was determined by a group of experts from the following federal agencies who sample different hazard/

product combinations annually: the Federal Veterinary Office (FVO), the Swiss Association of Cantonal Chemists, the Agroscope Institute for Food Science (IFS), the Federal Office of Public Health (FOPH), and the Federal Food Chain Unit. Hazard/product combinations were determined based on findings from the previous year.

These authors conducted a study on industry data collected through the NMPD between 2003 and 2010.[12] The study categorized cheese risk based on the type of production facility and style of cheese produced. The authors found that the type of production facility and cheese style were of greater significance than the pasteurization status of cheeses. The use of a risk-based approach identified risks significant to the consumer in order to protect public health, and it helped to ensure that regulations and guidelines were relevant and applicable to the products that were being produced. This example shows how the Swiss government identified a need within the cheese industry and gave the industry support to ensure that the industry was prepared for impending regulatory hurdles.

Instead of using sound science and education to help artisan cheesemakers, as was done in the Swiss approach, the FDA has instead used fear and intimidation and establishment of microbiological standards in the absence of scientific evaluation and risk assessment to provide requisite scientific evidence that these standards afforded public health protection.

CHAPTER 15

Advocacy

Imagine, just for a moment, that the FDA had been successful in its imposition of a ban on the use of wooden boards for aging cheese. The ripple effects throughout the artisan cheese industry would have been widely felt. In 2012, when I first heard of the FDA's intent to eliminate European cheesemaking practices, including use of wood, my thoughts immediately turned to Comté cheese and the vast implications for production of this magnificent product. Comté is an alpine-style cheese, made in the land of the Jura Massif region of France. In 2008, I had a chance to spend some time in that region while attending a wonderful wedding of two French cheesemakers. On my way to visit the Cave d'Affinage of Marcel Petite in Granges-Narboz, I stopped in the town of Arbois to visit the Maison de Louis Pasteur, the house of the famous scientist that sits in a remarkable state of preservation, complete with a home laboratory. In the laboratory were the swan-neck flasks used by Pasteur to disprove the theory of spontaneous generation. These experiments helped to save the French wine industry from demise. There were also ampules of materials that helped advance his work on development of vaccines. As I stood in the home and looked out the window, I could not help but think of the remarkable impact of Pasteur's work on foods of the region that use principles of fermentation to transform grapes and milk into the superb wines and cheeses that surround the Jura.

Comté is regarded as the King of Cheese in France, and Marcel Petite is one of the premier affineurs who helps each 70-pound wheel of cheese reach its full potential. As I toured the caves, the aging process was explained.

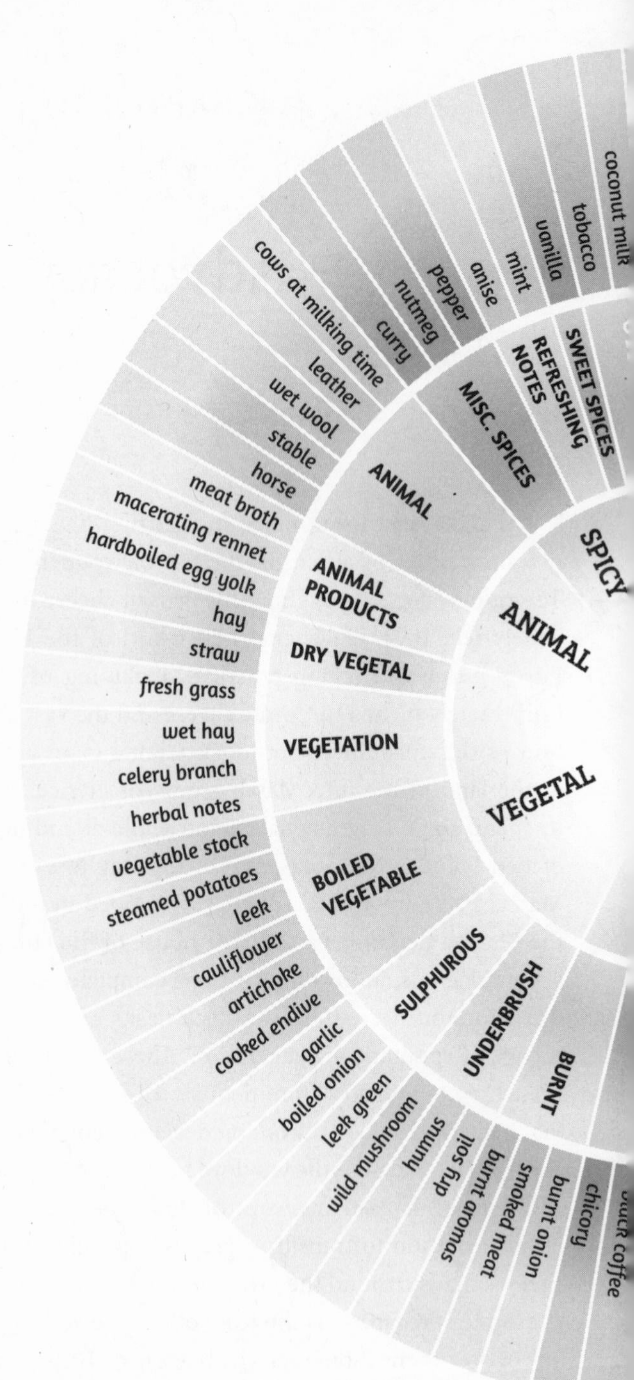

Figure 15.1. Comté sensory wheel. *Image courtesy CIGC*

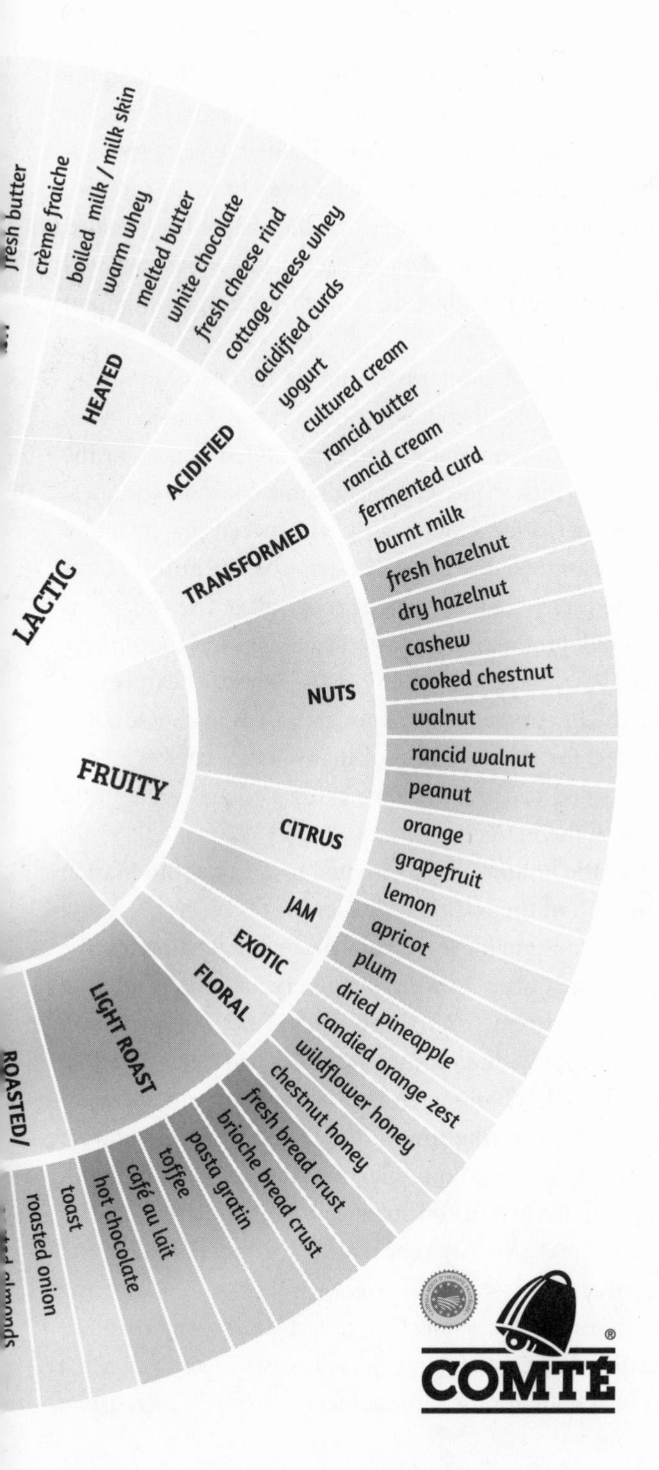
LACTIC
HEATED
ACIDIFIED
TRANSFORMED
fresh butter
crème fraiche
boiled milk / milk skin
warm whey
melted butter
white chocolate
fresh cheese rind
cottage cheese whey
acidified curds
yogurt
cultured cream
rancid butter
rancid cream
fermented curd
burnt milk
FRUITY
NUTS
CITRUS
JAM
EXOTIC
FLORAL
fresh hazelnut
dry hazelnut
cashew
cooked chestnut
walnut
rancid walnut
peanut
orange
grapefruit
lemon
apricot
plum
dried pineapple
candied orange zest
wildflower honey
chestnut honey
ROASTED/
LIGHT ROAST
fresh bread crust
brioche bread crust
pasta gratin
toffee
café au lait
hot chocolate
toast
roasted onion
COMTÉ

Just like apples in an orchard, not all reach maturation at exactly the same time; on a given day, some are ripe and others are not. The same is true for individual wheels of Comté made on the same day. Trained sensory specialists evaluate each wheel of cheese as it slowly ages over 10 to 14 months, checking for the sensory notes that indicate when the cheese has reached perfection. The Comté sensory wheel depicts each flavor note that was characterized by INRA (Institut National de la Recherche Agronomique) scientists working in the region.

Marcel Petite ripens much of France's Comté in Fort St. Antoine, a defunct military fort built in 1870 following the Franco-Prussian War. Today, over 100,000 wheels of Comté are aged on wooden boards in the underground cellars where thick stone walls maintain a constant and ideal temperature for the aging of Comté. Comté derives its special character and flavors from the milk of Montbéliarde cows that graze on the lush pastures full of dandelions in the Jura Massif. I thought about all of the enormous care that went into the milk production, the cheesemaking, and the *affinage lent* (slow aging) that allow the development of the luscious texture and flavors in this cheese. The FDA position on wooden boards in cheese aging shows a blatant disregard for this tradition of high-quality cheese manufacture. My thoughts turned to the incredible arrogance, ignorance, and disrespect associated with the FDA's intent.

In March 2010, I had the honor of being invited to speak at the Marian Koshland Science Museum of the National Academy of Sciences. The title of the presentation was a fun one for me: "Say Cheese: Understanding the Living Foods We Eat." At that time, Cowgirl Creamery had a Washington, DC–based store across the street from the Koshland Science Museum, so the lecture came complete with amazing cheeses. I had been asked by the American Society for Microbiology (ASM) if I could be interviewed prior to the lecture, as well as having the lecture taped. Chris Condayan, the manager for outreach in the communications department of ASM, conducted what became a lively and fun interview, which he turned into a public education video on ASM's MicrobeWorld. The video, *Cheese and Microbes* (MWV28, vimeo.com/4635585), went viral. The ASM, seeing tremendous public interest in the subject of cheese and a way to make microbiology interesting to scientists and nonscientists alike, asked if I would be willing to edit a book on the same subject. The book, *Cheese and*

Microbes, was published by the American Society for Microbiology in 2014.[1] The book included chapters from my favorite cheese scientists, including University of Vermont Professor Paul Kindstedt, who authored a chapter on how cheese is made, and Mother Noella Marcellino, the famous cheese nun who wrote about her work at the INRA in the Jura region of France through a Fulbright scholarship that helped her characterize the diversity of *Geotrichum candidum* strains, fungi that colonize nearly all surface-ripened cheeses during the early stages of cheese ripening. The Abbey of Regina Laudis, where Mother Noella resides, produces St. Nectaire cheese; and Mother Noella earned her PhD at the University of Connecticut in order to understand and improve the manufacture of St. Nectaire at the abbey. Dr. Eric Beuvier, also of the INRA, authored a chapter on the microbiology of hard and semi-hard mountain cheeses, such as Comté. As previously mentioned in earlier chapters of this book, Dr. Sylvie Lortal authored a chapter on the role of wooden tools as reservoirs of microbial biodiversity in cheesemaking. Drs. Rachel Dutton and Ben Wolfe authored a fascinating chapter that looked at an ecosystems approach to cheese microbiology. Dutton's work importantly recognized that microorganisms in cheese rinds function in highly ordered microbial communities. Her work, which applied cutting-edge molecular biology techniques using deep sequence DNA, found that cheesemaking technologies used throughout the world essentially select for the same ordered microbial communities as a function of the cheesemaking methods employed.

The ASM hosted a launch party of sorts for *Cheese and Microbes*, with Dr. Dutton and cheesemaker Mateo Kehler invited to give an evening presentation on "The Microbiology of Cheese" on June 10, 2014. The date of the event proved to be fortuitous, as it provided several of us the opportunity to use our time in Washington, DC, to speak with our congressional representatives about our concern for the FDA's proposed ban on the use of wooden boards in cheese aging. Our concerns were heard, and as a result, the Welch–Ryan amendment to the budget of the Department of Health and Human Services (HHS) was introduced on June 11, 2014, which prevented the FDA from using its funding to enforce a ban on wood in cheesemaking. This action is a prime example of advocacy. We need more efforts of this nature from scientists and cheesemakers alike if we want to protect our food system from regulatory overreach and abuse.

The adverse regulatory activity displayed by the US Food and Drug Administration toward the artisan cheese industry is an example of the broken food system that exists throughout the globe, and we urgently need to change it. As part of the World Economic Forum Annual Meeting in 2018, Corinna Hawkes, director of the Centre for Food Policy at the University of London, and Juergen Voegele, senior director of Food and Agriculture Global Practice for the World Bank, authored the article, "Our Food System Is Broken: Here Are 3 Ways to Fix It." These authors wrote:

> The food system is not a business and it has no CEO. Yet the facts stand for themselves: the food system is unequal, unsustainable, unstable, and in need of transformation. And it needs everyone—all stakeholders—to take action. . . .
>
> What is it that we need to do? The first thing is to radically change the way we think and act on food. It's no longer good enough to think that food is someone else's business or just to do our bit in isolation: we need to recognize that all our work affects food and figure out how to make changes in a connected way. This means recognizing that food is part of a "system." Once we start to see food as an interconnected system, from farm to fork, gate to plate, boat to throat, we begin to see how we all influence it in some way, and indeed, we are all affected by it.
>
> Second, let's see food as a solution. Food can help nourish people and planet, provide decent jobs, and provide enjoyment and happiness in our daily lives. A well-functioning food system provides a range of public benefits, from public health to the state of the environment, the economy, and society. Let's work harder to find solutions that recognise the interconnections in the system where we can achieve the greatest possible impact. . . .
>
> Third, to get this done, we must allocate our financial resources differently. Companies, governments, development agencies, philanthropists, and citizens alike. Are we putting our money into the most nutritious foods that contribute to a healthy diet? Are we using our tax dollars to support the people who produce nutritious food in ways that support resilient

> ecosystems? Are our investments maximising healthy diets while also enabling profitable business?[2]

Certainly, food regulations play a big role in this equation, and making sure that we have harmonized standards for food safety across the globe will do much to provide solutions to the inequities that currently exist. The regulatory reforms proposed in this book, such as having the USDA provide food safety oversight for all US-produced foods, afford many exciting opportunities to connect regulatory reform to rural economic development, environmental sustainability, and expansion of opportunities for small-scale businesses across not only the United States but throughout the globe. In France, aggressive measures to combat food waste, promote healthy lifestyles, and adopt eco-farming techniques have helped it top a ranking of nations based upon their food sustainability.[3]

Peril from evolving EU food regulations proved to be a motivator for reforms in France by small artisan cheesemakers whose livelihood was threatened by the war on raw milk cheese. The Fédération Régionale des Elevages de Côte d'Azur Alpes Provence (FRECAP; The Regional Federation of Animal Husbandry of the Cote D'Azur, the Alps, and Provence) was founded in 1981 to speak in a unified voice for the interests of many Provence-based cheesemakers. Among other things, FRECAP helped to define farmer-made cheese—*fromage fermier*—as cheese made under the following conditions: by a small family farm (one to three workers, 5 to 200 animals); with all the milk processed on the premises and coming from the animals raised and fed on the premises; with all cheeses produced and aged on the premises and sold by the selfsame farmer.[4] This group also lobbied successfully to have EU food laws rewritten to recognize the importance of traditional practices. In France, INRA scientists such as Dr. Marie-Christine Montel and others use strategies to promote beneficial bacteria in raw milk as the best way to combat microbial pathogens. Unfortunately, until recently, much of this information was only available in French. Through advocacy, Bronwen Percival of the Specialist Cheesemakers Association and Neal's Yard Dairy launched a campaign to translate a multiauthored book, *The Microbiology of Raw Milk*, into English. The original book, published in French in 2012 by a consortium of French scientists and coordinated by Cécile Laithier from the Institut d'Elevage, collected scientific information from researchers working to identify the benefits of raw milk cheese. This practical guide to raw milk microbiology is aimed at small-scale farmstead cheesemakers who are interested in promoting the natural microbial diversity of their milk, knowing that healthy and stable microbial communities contribute not only to cheese flavor but also to cheese safety.[5] Competitive exclusion is a microbiological strategy long used to ensure that beneficial organisms outnumber and outcompete bacterial pathogens.

In my own small Vermont community, I have witnessed the extraordinary ways in which food has completely transformed a small, economically struggling rural agricultural region. Yet much of the vibrancy in this community is threatened by regulatory activity that places intimidation and fear above education and science, and this must change. When we discuss changes to the food system, rarely do we discuss the degree to which regulations impact our food choices. The example of artisan cheesemakers and their battles with the FDA provides an important and illustrative case in point. Is the FDA best positioned to ensure the safety of cheese? My experience tells me no. Raw milk cheese made by licensed cheesemakers is safe cheese, as close attention is paid to the quality of the raw materials that are used in its production. Many of the foods that we process in the United States begin with materials that require a 5-log inactivation of bacterial pathogens to ensure they are safe. What if we instead produced foods that started with use of the highest-quality materials and ingredients? My laboratory at the University of Vermont has been conducting *Listeria* surveillance on farms producing milk used for artisan cheese production. A strategy that shows promise in reducing the potential for presence of *Listeria* in milk used for cheesemaking is to advise artisan cheese producers who make raw milk cheeses to eliminate silage feeding in favor of dry hay or pasture feeding. An excellent article by F. Driehuis and others reviews other microbiological hazards that can be transmitted to both animals and humans through silage feeding to dairy cattle.[6] Many PDO (Protected Designation of Origin) and AOP (Appellation d'Origine Protégée) European cheese varieties prohibit silage feeding for certain varieties of cheese due to the known microbiological hazards associated with this practice.

The FDA is a science-based agency and is required to use science as the basis for regulatory decisions. Yet food choice is impacted by economic, social, and environmental considerations. In Europe, these socioeconomic considerations are included in food safety debates and they influence regulatory approaches that govern food choice. Perhaps it is time for voices beyond science to impact our food choice here in the United States, and for us to become advocates for a different food system. I think the mainstream US dairy industry is missing an opportunity to embrace the production of world-class US artisan cheese. By targeting artisan cheese, regulatory activity is eliminating the growth of the dairy sector that could do so much to revitalize rural communities, eliminate concentrated animal feeding operations (CAFOs), reduce food waste and our carbon footprint, and produce foods that consumers in our nation of obsessed foodies want to consume, and in the process help small family farms stay in business. In 2012, 825 licensed, artisan cheesemakers in the United States produced more than 300 varieties of cheese, more than doubling the number from six years before.[7] US Retail dollar sales in the natural and specialty cheese sector are projected to reach $20.7 billion in 2020. Small-scale cheesemaking establishments in the United States represented almost half (46 percent) of all cheesemaking enterprises, an increase from 5 percent in 2007. As I tell my students at the University of Vermont, if you want to see this trend continue, vote with your pocketbook. Your local and highly educated cheesemonger will be more than happy to point you toward the newest and best artisan cheeses that the cheese counter has to offer.

On the topic of advocacy, I would be remiss if I did not pay tribute to my friend Cathy Strange, global cheese buyer for Whole Foods Market. Cathy has single-handedly done more for cheesemaker advocacy than any other person that I know. She travels the globe in pursuit of buying and importing the world's best cheeses. Her role at Whole Foods is critical to allowing market access through Whole Foods for cheeses from around the globe. She personally visits, supports, and advocates for great cheesemaking, and in the process has done much to improve the economies of rural communities in many countries. Her approach is unique and highly effective. In April 2017, Cathy organized a Whole Foods global sales meeting in the tiny town of Greensboro, Vermont, population 800. A new and magnificent Shakespearean theater had just been built and opened in Greensboro, and it provided the perfect backdrop for the meeting. At that meeting, Whole Foods presented Jason Hinds, a preeminent cheesemonger from Neal's Yard Dairy in London; Laure Dubouloz from Hervé Mons, one of France's preeminent affineurs; and Mateo Kehler from Jasper Hill—and all had their magnificent cheeses on display for tasting and evaluation. To me, this meeting spoke volumes about Cathy Strange and the values of the Whole Foods company. It would have been logistically easier to hold this meeting in a city (or really anywhere else on the planet, as anyone who has ever traveled to Greensboro knows). But Cathy made the commitment to the rural community where some world-class cheeses are being produced. This activity represents the kind of change in thinking about the food system espoused by Hawkes and Voegele.

If you love artisan cheese and want to support artisan cheesemakers, there are many ways for you, too, to become an advocate and to get involved. First, connect with your congressional representatives. Let them know that you are concerned about food safety, about FSMA and FDA regulations that have the potential to deliver a monolithic, industrialized food system at the expense of diverse and distributed small-scale food producers and artisan cheesemakers—who play an essential role in our rural economies and the sustainability of small farms. Your representatives desperately need to hear your voices to make these issues priorities.

If you want to stay abreast of issues that impact artisan cheese, subscribe to Janet Fletcher's *Planet Cheese* blog (www.janetfletcher.com/blog). As one of the best and most talented food writers in the United States, Janet can provide you with timely updates of what is new in the world of cheese, as well as regulatory activity that threatens our cheesemaking landscape and the availability of world-class cheese. If you want a deeper dive into the world of cheese, subscribe to *The Art of Eating* magazine (https://artofeating.com). This glorious publication edited and published by Edward Behr describes, in passionate and intellectual depth, cheeses, cheesemakers, and cheesemaking regions in all their glory. If you are a science nerd like me, subscribe to *Cheese Science Toolkit* (www.cheesescience.org). This blog, the brainchild of Pat Polowsky, a recent recipient of an MS degree at the University of Vermont, was the 2018 recipient of *Saveur* magazine's Food Obsessive Award, and won editor's choice for Best Single-Interest Blog. You could also subscribe to *Culture* magazine (https://culturecheesemag.com/go). This wonderful, globally focused publication delivers innovative stories about cheeses and cheesemakers and provides travel tips about cheese festivals here in the United States and abroad.

Another way to dive into the cheese world is to join Slow Food. This remarkable organization, based in Italy and founded by Carlo Petrini in 1986, advocates for raw milk cheesemakers working at the far reaches of the globe. As Slow Food states, "We are losing cheeses, animal breeds, pastures, herders, skills and ancient knowledge. It is not simply a question of the best milk and cheeses. Our food culture and the freedom to choose what we eat are at stake."[8] Closer to home, you can join Oldways (https://oldwayspt.org), a Boston-based food and nutrition nonprofit, dedicated to "Helping People Live Healthier, Happier Lives."[9] In October 2014, Oldways brought back the Cheese of Choice Coalition, which runs under the direction of Carlos Yescas. Oldways president, Sara Baer-Sinnott, stated, "Artisan, raw-milk, and traditional cheeses offer delicious tastes for consumers at home and in restaurants and provide a livelihood for passionate and dedicated cheese producers, distributors, importers and retailers worldwide. . . . To make sure that these cheeses continue to play such an important role, the Oldways Cheese Coalition will focus on protecting them from regulatory challenges and actively educate consumers and restaurateurs about their unique tastes and histories."

And finally, make sure you purchase and consume the great artisan cheeses of the United States and the rest of the globe. It is, of course, the best way to advocate for these products and pay tribute to the people who produce them.

GLOSSARY

ACoP: Assured Code of Practice (of the Specialist Cheesemakers Association)

ACS: American Cheese Society

ADPI: American Dairy Products Institute

Affinage: Process of aging and caring for cheese to encourage desired microbial growth and optimal flavor development

Aging: Holding cheeses in controlled environments for a specified time period

ALOP: Appropriate level of protection

AOC: Appellation d'Origine Contrôlée (French GI certification)

AOP: Appellation d'Origine Protégée (French GI certification)

APC: Aerobic plate count

ARS: Agricultural Research Service (USDA)

Artisan cheese: Cheeses produced on a small scale using traditional craftsmanship

ASM: American Society for Microbiology

Aw: Water activity

Bacteriophage: Viruses that attack bacterial starter cultures used in cheesemaking

CAFOs: Concentrated animal feeding operations

CCC: Cheese of Choice Coalition

CCFN: Consortium for Common Food Names

CCPs: Critical Control Points

CDC: Centers for Disease Control and Prevention

CFR: Code of Federal Regulations

CFSAN: Center for Food Safety and Applied Nutrition (FDA)

CFU: Colony forming unit; a term used for bacterial enumeration

Cheese cave: A facility used for cheese maturation

***Codex Alimentarius*:** Food Code; internationally recognized standards of practice used for food
CoRFiLaC: Consorzio per la Ricerca nel Settore della Filiera Lattiero Casearia (Sicily)
CPG: Compliance Policy Guide
CPS: Coagulase-positive staphylococci
Curd: Precipitation of protein (casein) from milk by enzymes or acid/temperature during cheesemaking
DDSV: Department of Veterinary Services Directorate (France)
DMI: Dairy Management Inc.
DOJ: US Department of Justice
EC: European Communities; incorporated into EU in 1993
***E. coli*:** *Escherichia coli*
EEC: Enteropathogenic *E. coli*
EU: European Union
FAO: Food and Agriculture Organization of the United Nations
Farmstead cheese: Cheese produced from milk collected on the same farm where the cheese is produced (also referred to as farmhouse cheese)
FDA: US Food and Drug Administration
FFDCA: Federal Food, Drug, and Cosmetic Act
FRECAP: Fédération Régionale des Elevages de Côte d'Azur Alpes Provence
FSA: Food Safety Administration
FSANZ: Food Standards Australia New Zealand
FSIS: Food Safety and Inspection Service (USDA)
FSMA: Food Safety Modernization Act (US)
FSRIA: Farm Security and Rural Investment Act
FSVP: Foreign Supplier Verification Program (part of FSMA)
FVO: Federal Veterinary Office (Switzerland)
GATT: General Agreement on Tariffs and Trade
GIs: Geographical indications
GMPs: Good manufacturing practices
HACCP: Hazard Analysis and Critical Control Points
HHS: US Department of Health and Human Services
HTST: High-temperature short-time pasteurization

ICMSF: International Commission on Microbiological Specification for Foods
IDF: International Dairy Federation
IDFA: International Dairy Foods Association
IFS: Agroscope Institute for Food Science (Switzerland)
IFST: Institute for Food Science and Technology (UK)
INRA: Institut National de la Recherche Agronomique (French National Institute for Agricultural Research)
LM: *Listeria monocytogenes*
Log: Logarithmic scale is a nonlinear scale used for a large range of positive multiples (that is, bacterial numbers)
Maître fromager: Master cheesemaker or affineur
MC: Microbiological criteria
MPN: Most probable number
NACMCF: National Advisory Committee on Microbiological Criteria for Foods
NAFTA: North American Free Trade Agreement
NASDA: National Association of State Departments of Agriculture
NMPD: National Monitoring Program of Milk and Dairy Products (Switzerland)
OAI: Official action indicated (FDA)
OIG: Office of Inspector General (HHS)
OMB: Office of Management and Budget
Pasteurization: Process applied to food to eliminate pathogens
PDO: Protected Designation of Origin
PEC: Pathogenic *E. coli*
PFGE: Pulsed-field gel electrophoresis
PGI: Protected Geographic Indication
pH: Potential hydrogen (a scale of acidity)
Queso fresco: A fresh Mexican-style soft cheese originally brought to Mexico from Spain
Raw milk cheese: Cheese made from unpasteurized milk
RTE: Ready-to-eat
SCA: Specialist Cheesemakers Association
SCC: Somatic cell count
Silage: Fermented hay or corn that is fed to ruminants

60-day rule: Holding requirement for cheese as an alternative to pasteurization
SMP: Salt in the moisture phase
SPC: Standard plate count
SPS: Sanitary and Phytosanitary Agreement
STEC: Shiga toxin–producing *E. coli*
Terroir: Taste and flavor imparted to a food from its natural environment
Thermization: Sub-pasteurization heat treatment of milk
Title 21: Portion of the Code of Federal Regulations that governs food and drugs within the United States
Tomme: Generic term referring to a group of cheeses commonly produced in the French Alps or Switzerland
TPC: Total plate count
TRIPS: Trade-Related Aspects of Intellectual Property Rights (TRIPS Agreement)
TSG: Traditional Speciality Guaranteed
T-TIP: Transatlantic Trade and Investment Partnership
UFPA: United Fresh Produce Association
U.S.C.: United States Code
USDA: US Department of Agriculture
USDEC: United States Dairy Export Council
USMCA: United States-Canada-Mexico Agreement
VIAC: Vermont Institute for Artisan Cheese
Washed rind: Also known as smear-ripened; cheeses are washed in brine solution to select for desired microbial species on the rind
WHO: World Health Organization
WTO: World Trade Organization

NOTES

Chapter 1: An Introduction to the Raw Milk Cheese Debate

1. Jeffrey P. Roberts, *The Atlas of American Farmstead Cheese: The Complete Guide to Making and Selling Artisan Cheeses* (White River Junction, VT: Chelsea Green Publishing, 2007).
2. Gregory McNeal, "FDA May Destroy American Artisan Cheese Industry," *Forbes*, June 9, 2014, https://www.forbes.com/sites/gregorymcneal/2014/06/09/fda-may-destroy-american-artisan-cheese-industry/#cde70863f6b8.
3. Burkhard Bilger, "Raw Faith," *New Yorker*, August 19, 2002, http://archives.newyorker.com/?i=2002-08-19#folio=152.
4. David Schrieberg,"Why Your Genuine French Camembert Cheese Is in Danger," *Forbes*, February 25, 2018, https://www.forbes.com/sites/davidschrieberg1/2018/02/25/why-your-genuine-french-camembert-cheese-is-in-danger/#112ec3b11545.
5. Larissa Zimberoff, "One of the World's Great Cheeses Might Be Going Extinct," *Bloomberg*, June 13, 2017, https://www.bloomberg.com/news/articles/2017-06-13/camembert-cheese-might-be-going-extinct.
6. D. Corroler et al., "An Ecological Study of Lactococci Isolated from Raw Milk in the Camembert Cheese Registered Designation of Origin Area," *Applied and Environmental Microbiology* 64, no. 12 (December 1998): 4729–35.
7. James Brookes, "Charles Highlights Plight of French Cheeses During Paris Visit," *Royal Central*, November 30, 2015, http://royalcentral.co.uk/uk/charlesandcamilla/charles-highlights-plight-of-french-cheeses-during-paris-visit-56378.

Chapter 2: The Historical Context for Raw Milk Cheesemaking

1. "Lucy Appleby," Obituaries, *Telegraph*, May 2, 2008, http://www.telegraph.co.uk/news/obituaries/1921997/Lucy-Appleby.html.

2. Paul Kindstedt, "Trade (Historical)," in *The Oxford Companion to Cheese*, ed. Catherine W. Donnelly (New York: Oxford University Press, 2016), 724.
3. David W. Fleming et al., "Pasteurized Milk as a Vehicle of Infection in an Outbreak of Listeriosis," *New England Journal of Medicine* 312, no. 7 (February 1985): 404–7, http://doi.org/10.1056/NEJM198502143120704.
4. Megan Slack, "Behind the Scenes at the France State Dinner: See the Menu," *Obama White House Blog*, February 11, 2014, https://obamawhitehouse.archives.gov/blog/2014/02/11/behind-scenes-france-state-dinner-see-menu.
5. Paul Kindstedt, "Cheesemaking in the New World: The American Experience," in *American Farmstead Cheese*, ed. Paul Kindstedt with the Vermont Cheese Council (White River Junction, VT: Chelsea Green, 2005), 17–35.
6. E. A. Zottola and L. B. Smith, "Pathogens in Cheese," *Food Microbiology* 8, no. 3 (September 1991): 171–82, https://doi.org/10.1016/0740-0020(91)90048-7.
7. US Food and Drug Administration, "Part 19-Cheeses; Processed Cheeses; Cheese Foods; Cheese Spreads, and Related Foods: Definition and Standards of Identity [Docket no. FDC-46]. Final Rule," *Federal Register*, August 24, 1950, 5656–90.

Chapter 3: Is Raw Milk Cheese Really Risky?

1. Eric A. Johnson et al., "Microbiological Safety of Cheese Made from Heat-Treated Milk, Part I. Executive Summary, Introduction and History," *Journal of Food Protection* 53, no. 5 (May 1990): 441–52, https://doi.org/10.4315/0362-028X-53.5.441; Eric A. Johnson et al., "Microbiological Safety of Cheese Made from Heat-Treated Milk, Part II. Microbiology," *Journal of Food Protection* 53, no. 6 (June 1990): 519–40, https://doi.org/10.4315/0362-028X-53.6.519.
2. Sean F. Altekruse et al., "Cheese-Associated Outbreaks of Human Illness in the United States, 1973 to 1992: Sanitary Manufacturing Practices Protect Consumers," *Journal of Food Protection* 61, no. 10 (October 1998): 1405–7, https://doi.org/10.4315/0362-028X-61.10.1405.
3. Andrew J. Knight et al., "Listeria in Raw Milk Soft Cheese: A Case Study of Risk Governance in the United States Using the IRGC Framework," in *Global Risk Governance, International Risk Governance Council*

Bookseries, vol 1, eds. O. Renn and K. D.Walker (Dordrecht, NL: Springer, 2008), 179–220.

4. Beth P. Bell et al., "A Multistate Outbreak of Escherichia coli O157:H7—Associated Bloody Diarrhea and Hemolytic Uremic Syndrome from Hamburgers: The Washington Experience," *Journal of the American Medical Association* 272, no.17 (November 1994): 1349–53, http://doi.org/10.1001/jama.1994.03520170059036.
5. Christine J. Reitsma and David R. Henning, "Survival of Enterohemorrhagic *Escherichia coli* O157:H7 During the Manufacture and Curing of Cheddar Cheese," *Journal of Food Protection* 59, no. 5 (May 1996): 460–64, http://doi.org/10.4315/0362-028X-59.5.460.
6. J. E. Schlesser et al., "Survival of a Five-Strain Cocktail of Escherichia coli O157:H7 During the 60-Day Aging Period of Cheddar Cheese Made from Unpasteurized Milk," *Journal of Food Protection* 69, no. 5 (May 2006): 990–98, https://doi.org/10.4315/0362-028X-69.5.990.
7. Christophe J. Büla, Jacques Bille, and Michel P. Glauser, "An Epidemic of Food-Borne Listeriosis in Western Switzerland: Description of 57 Cases Involving Adults," *Clinical Infectious Diseases* 20, no. 1 (January 1995): 66–72, http://doi.org/10.1093/clinids/20.1.66.
8. Michael J. Linnan et al., "Epidemic Listeriosis Associated with Mexican-Style Cheese," *New England Journal of Medicine* 319, no. 13 (September 1988): 823–38, http://doi.org/10.1056/NEJM198809293191303.
9. Melanie Rudolf and Siegfried Scherer, "High Incidence of *Listeria monocytogenes* in European Red Smear Cheese," *International Journal of Food Microbiology* 63, no. 1–2 (January 2001): 91–98, http://doi.org/10.1016/S0168-1605(00)00413-X.
10. David J. Hornsby, "Multilevel Governance in Domestic Regulatory Conflict: Raw-Milk Cheese in Canada" (Paper presented at the Canadian Political Science Association Annual Meeting, Waterloo, Canada, May 18, 2011), https://www.cpsa-acsp.ca/papers-2011/Hornsby.pdf.
11. Hornsby, "Multilevel Governance."
12. Institute of Medicine and National Research Council, "Scientific Criteria to Ensure Safe Food" (Washington, DC: National Academies Press, 2003).

13. US Food and Drug Administration, "Understanding Potential Intervention Measures to Reduce the Risk of Foodborne Illness from Consumption of Cheese Manufactured from Unpasteurized Milk," *Federal Register*, August 3, 2015, 80 FR 46023, Docket No. FDA-2015-N-2596, https://www.federalregister.gov/documents/2015/08/03/2015-18972/understanding-potential-intervention-measures-to-reduce-the-risk-of-foodborne-illness-from.
14. Adam J. Langer et al., "Nonpasteurized Dairy Products, Disease Outbreaks, and State Laws—United States, 1993–2006," *Emerging Infectious Diseases* 18, no. 3 (March 2012): 385–91, http://doi.org/10.3201/eid1803.111370.
15. Marcia L. Headrick et al., "The Epidemiology of Raw Milk-Associated Foodborne Disease Outbreaks Reported in the United States, 1973 through 1992," *American Journal of Public Health* 88, no. 8 (August 1998): 1219–21, http://doi.org/10.2105/AJPH.88.8.1219.
16. Sara H. Cody et al., "Two Outbreaks of Multidrug-Resistant *Salmonella* Serotype Typhimurium DT104 Infections Linked to Raw-Milk Cheese in Northern California," *Journal of the American Medical Association* 281, no. 19 (May 1999): 1805–10, http://doi.org/10.1001/jama.281.19.1805.
17. C. Méndez Martínez et al., "Brucellosis Outbreak Due to Unpasteurized Raw Goat Cheese in Andalucia (Spain), January - March 2002," *Eurosurveillance* 8, no. 7 (July 2003): 164–68.
18. L. Hannah Gould, Elisabeth Mungai, and Casey Barton Behravesh, "Outbreaks Attributed to Cheese: Differences Between Outbreaks Caused by Unpasteurized and Pasteurized Dairy Products, United States, 1998–2011," *Foodborne Pathogens and Disease* 11, no. 7 (June 2014): 545–51, http://doi.org/10.1089/fpd.2013.1650.
19. Food Standards Australia New Zealand, *Microbiological Risk Assessment of Raw Milk Cheese*, Canberra, December 2009, http://www.foodstandards.gov.au/code/proposals/documents/P1007%20PPPS%20for%20raw%20milk%201AR%20SD3%20Cheese%20Risk%20Assessment.pdf.
20. K. E. Heiman et al., "Multistate Outbreak of Listeriosis Caused by Imported Cheese and Evidence of Cross-Contamination of Other

Cheeses, USA, 2012," *Epidemiology and Infection* 144, no.13 (2016): 2698–708, http://doi.org/10.1017/S095026881500117X.

21. Dennis J. D'Amico and Catherine W. Donnelly, "FDA's Domestic and Imported Cheese Compliance Program Results: January 1, 2004–December 31, 2006," *Food Protection Trends* 31, no. 4 (2011): 216–26.
22. US Food and Drug Administration, "Joint FDA / Health Canada Quantitative Assessment of the Risk of Listeriosis from Soft-Ripened Cheese Consumption in the United States and Canada," November 2017, https://www.fda.gov/food/foodscienceresearch/risksafety assessment/ucm429410.htm.

Chapter 4: Why Raw Milk Cheese Is Not Raw

1. H. P. Bachmann and U. Spahr, "The Fate of Potentially Pathogenic Bacteria in Swiss Hard and Semihard Cheeses Made from Raw Milk," *Journal of Dairy Science* 78, no. 3 (March 1995): 476–83, http://doi .org/10.3168/jds.S0022-0302(95)76657-7.
2. Luisa Pellegrino and P. Resmini, "Cheesemaking Conditions and Compositive Characteristics Supporting the Safety of the Raw Milk Cheese Italian Grana," *Scienza e Tecnica Lattiero Casearia* 52 (2001): 105–14.
3. G. Panari et al., "Il Comportamento dei Batteri Potenzialmente Patogeni nella Produzione di ParmigianoReggiano [The Behavior of Potentially Pathogenic Bacteria in the Production of Parmigiano-Reggiano]," *Scienza e Tecnica Lattiero Casearia* 55 (2004): 137–46.
4. World Trade Organization, "Extracts from SPS Committee Meeting Summary Reports," *Sanitary and Phytosanitary Information Management System*, http://spsims.wto.org/en/SpecificTradeConcerns/View/29.
5. Food Standards Australia New Zealand, *Final Assessment Report. Proposal P263. Safety Assessment of Raw Milk Very Hard Cooked-Curd Cheeses*, Canberra, November 20, 2002, http://www.foodstandards .gov.au/standardsdevelopment/proposals/proposalp263hardraw milkcheese/index.cfm. Food Standards Australia New Zealand, *Review of the Decision to Exempt Raw Milk Extra Hard Grating Cheeses from the Milk Heat Treatment Requirement in Standard 1.6.2 of the Food Standards Code—Discussion paper*, Canberra, July 22, 2003.

6. Food Standards Australia New Zealand, *FSANZ Guidelines Determining the Equivalence of Food Safety Measures*, 2004, http://www.foodstandards.gov.au/publications/documents/Equivalence%20Determination%20 Guidelines_jan04.pdf.
7. World Trade Organization, "Extracts from SPS Committee Meeting."
8. Food Standards Australia New Zealand, "Application A499 - To permit the Sale of Roquefort Cheese," 2005, http://www.foodstandards.gov.au/code/applications/pages/applicationa499toper2374.aspx.

Chapter 5: EU versus US Regulations

1. World Trade Organization, *The General Agreement of Tarriffs and Trade 1947*, https://www.wto.org/english/docs_e/legal_e/gatt47_01_e.htm.
2. World Trade Organization, *Understanding the Agreement on Sanitary and Phytosanitary Measures*, 1998, https://www.wto.org/english/tratop_e/sps_e/spsund_e.htm.
3. World Trade Organization, "What is the WTO?," https://www.wto.org/english/thewto_e/whatis_e/whatis_e.htm.
4. World Trade Organization, "The WTO and the FAO/WHO Codex Alimentarius," https://www.wto.org/english/thewto_e/coher_e/wto_codex_e.htm.
5. Codex Alimentarius Commission, "Joint FAO/WHO Food Standards Programme," Geneva, June 23–28, 1997, http://www.fao.org/3/w3700e/w3700e00.htm.
6. Hornsby, "Multilevel Governance."
7. Véronique Lafarge et al., "Raw Cow Milk Bacterial Population Shifts Attributable to Refrigeration," *Applied and Environmental Microbiology* 70, no. 9 (September 2004): 5644–50, http://doi.org/10.1128/AEM.70.9.5644-5650.2004.
8. Dennis J. D'Amico, Errol Groves, and Catherine W. Donnelly, "Low Incidence of Foodborne Pathogens of Concern in Raw Milk Utilized for Farmstead Cheese Production," *Journal of Food Protection* 71, no. 8 (2008): 1580–89.
9. Dennis J. D'Amico and Catherine W. Donnelly, "Microbiological Quality of Raw Milk Used for Small-Scale Artisan Cheese Production in Vermont: Effect of Farm Characteristics and Practices," *Journal of Dairy Science* 93, no. 1 (January 2010): 134–47, http://doi.org/10.3168/jds.2009-2426.

Chapter 6: Redefining Pasteurization

1. National Advisory Committee on the Microbiological Criteria for Foods, "Requisite Scientific Parameters for Establishing the Equivalence of Alternative Methods of Pasteurization," *Journal of Food Protection* 69, no. 5 (May 2006): 1190–1216, https://doi.org/10.4315/0362-028X-69.5.1190.
2. National Advisory Committee, "Requisite Scientific Parameters."
3. Food and Agriculture Organizations of the United Nations, "Code of Hygienic Practice for Milk and Milk Products," 2004, http://www.fao.org/fileadmin/user_upload/livestockgov/documents/CXP_057e.pdf.

Chapter 7: FDA's Assault on Artisan Cheese, Phase 1

1. Dennis J. D'Amico and Catherine W. Donnelly, "FDA's Domestic and Imported Cheese Compliance Program Results: January 1, 2004–December 31, 2006," *Food Protection Trends* 31, no. 4 (2011): 216–26.
2. Michael A. Chappell, "Draft Compliance Policy Guide Sec. 527.300 Dairy Products—Microbial Contaminants and Alkaline Phosphatase Activity (CPG 7106.08)," *Federal Register* 229, no. 74 (December 1, 2009): 62795, http://edocket.access.gpo.gov/2009/pdf/E9-28756.pdf.
3. "Compliance Policy Guide Sec. 527.300 Dairy Products—Microbial Contaminants and Alkaline Phosphatase Activity," *Federal Register* notice of availability, December 23, 2010.
4. "'Confusion' Over FDA Rules Could Endanger Your Favorite French Cheese," *Q13FOX*, September 8, 2014, https://q13fox.com/2014/09/08/confusion-over-fda-rules-could-endanger-your-favorite-french-cheese/.
5. International Commission of Microbiological Specifications of Foods (ICMSF), *Microorganisms in Foods Book 2: Use of Data for Assessing Process Control and Product Acceptance* (Toronto: University of Toronto Press, 1986), 168–169, http://www.icmsf.org/publications/books.
6. National Advisory Committee on Microbiological Criteria for Foods, "Response to Questions Posed by the Department of Defense Regarding Microbiological Criteria as Indicators of Process Control or Insanitary Conditions," *Journal of Food Protection* 81, no. 1 (January 2018): 115–41, https://doi.org/10.4315/0362-028X.JFP-17-294.
7. US Food and Drug Administration, "Microbiological Surveillance Sampling: FY14-16 Raw Milk Cheese Aged 60 Days," updated February

26, 2018, https://www.fda.gov/Food/ComplianceEnforcement/Sampling/ucm510799.htm.

Chapter 8: FDA's Assault on Artisan Cheese, Phase 2

1. Bénédicte Coudé and Bill Wendorff, "Future Uses of Wooden Boards for Aging Cheeses," *Dairy Pipeline* 25, no. 1 (2013), https://www.cdr.wisc.edu/sites/default/files/insider/resources/wooden_boards.pdf.
2. Sylvie Lortal, Giuseppe Licitra, and Florence Valence, "Wooden Tools: Reservoirs of Microbial Biodiversity in Traditional Cheesemaking," *Microbiology Spectrum* 2, no. 1 (January 2014), http://doi.org/10.1128/microbiolspec.CM-0008-2012.
3. US Food and Drug Administration, "Warning Letter NYK-2013-1," Inspections, Compliance, Enforcement, and Criminal Investigations, October 23, 2012, http://wayback.archive-it.org/7993/20170112191456/http://www.fda.gov/ICECI/EnforcementActions/WarningLetters/2012/ucm325714.htm.
4. US Food and Drug Administration, "United States Enters Consent Decree with New York Cheese Producer Due to Listeria Contamination," press release, April 29, 2014, http://www.fda.gov/NewsEvents/Newsroom/PressAnnouncements/ucm395339.htm.
5. C. Mariani et al., "Biofilm Ecology of Wooden Shelves Used in Ripening the French Raw Milk Smear Cheese Reblochon de Savoie," *Journal of Dairy Science* 90, no. 4 (April 2007): 1653–61, https://doi.org/10.3168/jds.2006-190.
6. Peter Zangerl et al., "Survival of *Listeria monocytogenes* After Cleaning and Sanitation of Wooden Shelves Used for Cheese Ripening," *European Journal of Wood and Wood Products* 68, no. 4 (November 2010): 415–19, http://doi.org/10.1007/s00107-009-0381-6.
7. Food Standards Australia New Zealand, "Application A499 - To permit the Sale of Roquefort Cheese," 2005, http://www.foodstandards.gov.au/code/applications/pages/applicationa499toper2374.aspx.
8. Jeanne Carpenter, "Game Changer: FDA Rules No Wooden Boards in Cheese Aging," *Cheese Underground* (blog), June 7, 2014, https://cheeseunderground.com/2014/06/07/game-changer-fda-rules-no-wooden-boards-in-cheese-aging/.

9. Gregory S. McNeal, "FDA Backs Down in Fight over Aged Cheese," *Forbes*, June 10, 2014, https://www.forbes.com/sites/gregorymcneal/2014/06/10/fda-backs-down-in-fight-over-aged-cheese.

Chapter 9: *Listeria* and the Soft Cheese Risk Assessment

1. Catherine W. Donnelly, "*Listeria monocytogenes*: A Continuing Challenge," *Nutrition Reviews* 59, no. 6 (June 2001): 183–94, https://doi.org/10.1111/j.1753-4887.2001.tb07011.x.
2. Veronique Goulet et al., "Listeriosis from Consumption of Raw-Milk Cheese," *Lancet* 345, no. 8964 (June 1995): 1581–82, https://doi.org/10.5555/uri:pii:S0140673695911359.
3. Centers for Disease Control and Prevention, "Outbreak of Listeriosis Associated with Homemade Mexican-Style Cheese—North Carolina, October 2000–January 2001," *Morbidity and Mortality Weekly Report* 50, no. 26 (July 2001): 560–62.
4. Kathrin Rychli et al., "Genome Sequencing of *Listeria monocytogenes* 'Quargel' Listeriosis Outbreak Strains Reveals Two Different Strains with Distinct *In Vitro* Virulence Potential," *PLoS ONE* 9, no. 2 (February 2014): e89964, http://doi.org/10.1371/journal.pone.0089964.
5. Dennis J. D'Amico and Catherine W. Donnelly, "Microbiological Quality of Raw Milk Used for Small-Scale Artisan Cheese Production in Vermont: Effect of Farm Characteristics and Practices," *Journal of Dairy Science* 93, no. 1 (January 2010): 134–47, http://doi.org/10.3168/jds.2009-2426.
6. Department of Health and Human Services, memorandum, May 25, 2012, https://www.cheesesociety.org/wp-content/uploads/2016/06/Environmental-Listeria-Sampling-at-Soft-Cheese-Firms.pdf.
7. The Innovation Center for U.S. Dairy, "About Us," https://www.usdairy.com/about-us/about-the-innovation-center.

Chapter 10: Geographical Indications

1. Council of the European Union, "Council Regulation (EEC) No 2081/92 of 14 July 1992 on the Protection of Geographical Indications and Designations of Origin for Agricultural Products and Foodstuffs," July 14, 1992, https://publications.europa.eu/en/publication-detail/-/publication/7332311d-d47d-4d9b-927e-d953fbe79685.

2. Dick Groves, "Dairy Trade Might Not Get Any Freer, Or Fairer, For A While," *Cheese Reporter*, 2016, http://www.cheesereporter.com/archive/2016editorial.htm.
3. C. Hough, "The EU Tries to Grab All the Cheese," *Politico*, June 8, 2016, https://www.politico.com/agenda/story/2016/06/european-union-trade-cheese-geographical-indicators-000141.
4. World Trade Organization, "DS 174 European Communities—Protection of Trademarks and Geographical Indications for Agricultural Products and Foodstuffs," April 21, 2006, https://www.wto.org/english/tratop_e/dispu_e/cases_e/ds174_e.htm.
5. US Food and Drug Administration, "Code of Federal Regulations Title 21 Volume 2 Chapter 1-Food and Drug Administration, Department of Health and Human Serivces. Subchapter B-Food for Human Consumption Part 133 Cheeses and Related Cheese Products Subpart B—Requirements for Specific Standardized Cheese and Related Products Sec. 133.165 Parmesan and Reggiano Cheese," April 1, 2018, https://www.accessdata.fda.gov/scripts/cdrh/cfdocs/cfcfr/cfrsearch.cfm?fr=133.165.
6. Consortium of Parmigiano Reggiano Cheese, "Specification and Regulations," 2011, https://www.parmigianoreggiano.com/consortium/rules_regulation_2/default.aspx.
7. Consortium of Parmigiano Reggiano Cheese, "Specification and Regulations."
8. Larry Olmsted, "Most Parmesan Cheeses In America Are Fake, Here's Why," *Forbes*, November 19, 2012, https://www.forbes.com/sites/larryolmsted/2012/11/19/the-dark-side-of-parmesan-cheese-what-you-dont-know-might-hurt-you/#258c158e4645.
9. Jason Wilson, "How Good Has U.S. Cheese Become? Good Enough to Worry the Italians!," *Washington Post*, October 3, 2017, https://wisconsindairybuzz.com/2017/10/03/the-washington-post-how-good-has-u-s-cheese-become-good-enough-to-worry-the-italians/.
10. M. Corrado, "Italian Perspective on the Importance of Geographical Indications and Protected Designations of Origin Status for Parmigiano-Reggiano Cheese," *Chicago Kent Journal of Intellectual Property* 16, no. 2 (2017), https://scholarship.kentlaw.iit.edu/cgi/view content.cgi?article=1182&context=ckjip.

11. US Patent and Trademark Office, "Summary of the Report of the Panel (WT/DS174/R) of March 15, 2005 Regarding the Complaint by the United States against the European Communities on the Protection of Trademarks and Geographical Indications for Agricultural Products and Foodstuffs," 2005, https://www.uspto.gov/sites/default/files/web/offices/dcom/olia/globalip/pdf/case_summary.pdf.
12. Official Journal of the European Union, "Regulation (EU) No 1151/2012 of the European Parliament and of the Council of 21 November 2012 on Quality Schemes for Agricultural Products and Foodstuffs," Volume 55 (December 14, 2012): OJ L 343, 14.12.2012, p. 1.
13. Official Journal of the European Union, "Council Regulation (EC) No 510/2006 of 20 March 2006 on the Protection of Geographical Indications and Designations of Origin for Agricultural Products and Foodstuffs," Volume 49 (March 31, 2006): OJ L 93, 31.3.2006, p. 12; Official Journal of the European Union, "Council Regulation (EC) No 509/2006 of 20 March 2006 on Agricultural Products and Foodstuffs as Traditional Specialities Guaranteed," Volume 49 (March 31, 2006): OJ L 93, 31.3.2006, p. 1.
14. K. W. Watson, "Reign of *Terroir*: How to Resist Europe's Efforts to Control Common Food Names as Geographical Indications," Policy Analysis No. 787. Cato Institute. February 16, 2016, https://www.cato.org/publications/policy-analysis/reign-terroir-how-resist-europes-efforts-control-common-food-names#full.
15. Consortium for Common Food Names, http://www.commonfoodnames.com/; U.S. Dairy Export Council, "Cheese Variety and Affordability," http://www.thinkusadairy.org/products/cheese/cheese-variety-and-affordability.
16. The Hagstrom Report, "US Dairy Criticizes GI Provision in EU-Mexico Agreement, EU Farmers Praise It," *Fence Post*, April 26, 2018, https://www.thefencepost.com/news/us-dairy-criticizes-gi-provision-in-eu-mexico-agreement-eu-farmers-praise-it/.
17. Michael Handler, "The WTO Geographical Indications Dispute," *Modern Law Review* 69 (2006): 70–80.
18. Consortium for Common Food Names, "USMCA Breaks New Ground with Better GI Policy as Defending Common Names Becomes a U.S. Priority," October 8, 2018, http://www.commonfoodnames.com

/usmca-breaks-new-ground-with-better-gi-policy-as-defending
-common-names-becomes-a-u-s-priority/.
19. Consortium for Common Food Names, "USMCA Breaks New Ground."
20. EURACTIV.com with Reuters, "US Senators Shocked by EU's Cheese-Name Claims," March 12, 2014, https://www.euractiv.com/section
/agriculture-food/news/us-senators-shocked-by-eu-s-cheese-name
-claims/.
21. Barbara Boland, "Former Employee of 'Big Cheese' Wrote FDA Letter that Put NY Artisan Cheese Makers Out of Work," CNSNews, June 13, 2014, https://www.cnsnews.com/mrctv-blog/barbara-boland
/former-employee-big-cheese-wrote-fda-letter-put-ny-artisan
-cheese-makers.

Chapter 11: The Food Safety Modernization Act

1. Elaine Scallan et al., "Foodborne Illness Acquired in the United States—Major Pathogens," *Emerging Infectious Diseases* 17, no. 1 (2011): 7–15.
2. Centers for Disease Control and Prevention, "Multistate Outbreak of Salmonella Serotype Tennessee Infections Associated with Peanut Butter—United States, 2006–2007," *MMWR* 56 no. 21 (2007): 521–24, https://www.cdc.gov/mmwr/preview/mmwrhtml/mm5621a1.htm.
3. Juliana Grant et al., "Spinach-Associated *Escherichia coli* O157:H7 Outbreak, Utah and New Mexico, 2006," *Emerging Infectious Diseases* 14, no. 10 (2008): 1633–36.
4. Centers for Disease Control and Prevention, "Outbreak of *Listeria monocytogenes* Infections Associated with Pasteurized Milk from a Local Dairy—Massachusetts, 2007," *MMWR* 57, no. 40 (2008): 1097–1100.
5. Centers for Disease Control and Prevention, "Surveillance for Foodborne Disease Outbreaks—United States, 2007," *MMWR* 59, no. 31 (2010): 973–79.
6. H.R. 2749 (111th), Food Safety Enhancement Act of 2009, https://www
.govtrack.us/congress/bills/111/hr2749.
7. S. 510 - FDA Food Safety Modernization Act, https://www.congress.gov
/bill/111th-congress/senate-bill/510.
8. Federal Legislation 111th Congress 2009–2010, https://farmtoconsumer
.org/federal/111_Cong-S510.htm.

9. Organic Consumers Association Info on S510—The Food Safety Modernization Act, https://www.organicconsumers.org/news/info-s510-food-safety-modernization-act.
10. US Food and Drug Administration FDA Food Safety Modernization Act (FSMA), https://www.fda.gov/food/guidanceregulation/fsma/.
11. Michael R. Taylor and Howard R. Sklamberg, "Internationalizing Food Safety: FDA's Role in the Global Food System," *Harvard International Review* 37, no. 3 (2016): 32–37.
12. US Food and Drug Administration 2019, Investigation Summary: Factors Potentially Contributing to the Contamination of Romaine Lettuce Implicated in the Fall 2018 Multi-State Outbreak of *E. coli* O157:H7 February 13, https://www.fda.gov/Food/RecallsOutbreaksEmergencies/Outbreaks/ucm631243.htm.
13. Centers for Disease Control and Prevention 2007, Investigation of an *Escherichia coli* 0157:H7 Outbreak Associated with Dole Pre-Packaged Spinach Jan. 12, https://www.cdc.gov/nceh/ehs/docs/investigation_of_an_e_coli_outbreak_associated_with_dole_pre-packaged_spinach.pdf.
14. Centers for Disease Control and Prevention PulseNet, https://www.cdc.gov/pulsenet/index.html.
15. US Food and Drug Administration GenomeTrakr Network, https://www.fda.gov/food/foodscienceresearch/wholegenomesequencingprogramwgs/ucm363134.htm.
16. State of California—Health and Human Services Agency Department of Public Health, 2007, "Investigation of an Escherichia coli O157:H7 Outbreak Associated with Dole Pre-Packaged Spinach March 21," https://www.marlerblog.com/files/2013/02/2006_Spinach_Report_Final_01.pdf.
17. Casey Barton Behravesh et al., "2008 Outbreak of Salmonella Saintpaul Infections Associated with Raw Produce," *New England Journal of Medicine* 364, no. 10 (2011): 918–27.
18. Pew Charitable Trusts, "Lessons To Be Learned from the 2008 *Salmonella* Saintpaul Outbreak," November 17, 2008, https://www.pewtrusts.org/~/media/legacy/uploadedfiles/phg/content_level_pages/reports/psprptlessonssalmonella2008pdf.pdf.
19. Food and Water Watch, "FDA Approves Irradiation Despite Uncertainties About Consumer Safety," Statement by Wenonah Hauter,

Food & Water Watch Executive Director, April 21, 2008, https://www.foodandwaterwatch.org/news/fda-approves-irradiation-despite-uncertainties-about-consumer-safety.
20. Daniel R. Levinson, "FDA Inspections of Domestic Food Facilities," Department of Health and Human Services Office of Inspector General, April (2010), https://oig.hhs.gov/oei/reports/oei-02-08-00080.pdf.
21. Daniel R. Levinson, "Challenges Remain in FDA's Inspections of Domestic Food Facilities," Department of Health and Human Services Office of Inspector General, September 2017, https://oig.hhs.gov/oei/reports/oei-02-14-00420.pdf.
22. Ana Allende and James Monaghan, "Irrigation Water Quality for Leafy Crops: A Perspective of Risks and Potential Solutions," *International Journal of Environmental Research and Public Health* 12, no. 7 (2015): 7457–77.
23. National Association of State Departments of Agriculture,"NASDA Announces Cooperative Agreement to Implement National Produce Safety Rule Press Release," September 16, 2014, https://www.nasda.org/news/nasda-announces-cooperative-agreement-to-implement-national-produce-safety-rule.
24. US Food and Drug Administration, "FDA Announces Cooperative Agreement to Implement National Produce Safety Rule," September 16, 2014, http://wayback.archive-it.org/7993/20171114120224/https://www.fda.gov/Food/NewsEvents/ConstituentUpdates/ucm414777.htm.
25. National Association of State Departments of Agriculture, "NASDA Announces Cooperative Agreement to Implement National Produce Safety Rule."
26. James Andrews, "Harvard Law School to Host Conference on Food Safety Law," December 10, 2013, https://www.foodsafetynews.com/2013/12/harvard-law-school-to-host-conference-on-food-safety-law/.
27. Shawn Stevens, *FDA's War on Pathogens: Criminal Charges for Food Company Executives and Quality Assurance Managers*, Food Industry Counsel LLC, https://www.foodindustrycounsel.com/blog/white-paper-fdas-war-on-pathogens.
28. Centers for Disease Control and Prevention, "Multistate Outbreak of Listeriosis Linked to Blue Bell Creameries Products (Final Update)," 2015, https://www.cdc.gov/listeria/outbreaks/ice-cream-03-15/index.html.

Chapter 12: Timeline of Key Events and Hard Questions

1. American Cheese Society, "American Cheese Society Response to U.S. FDA Request for Information Regarding Safe Production of Cheese from Unpasteurized Milk," October 29, 2015, https://www.cheesesociety.org/wp-content/uploads/2015/10/ACS-Comments-to-FDA-2015-N-2596.pdf.
2. US Food and Drug Administration,"FY 2014–2016 Microbiological Sampling Assignment Summary Report: Raw Milk Cheese Aged 60 Days," Office of Compliance Center for Food Safety and Applied Nutrition, July 21, 2016, https://www.fda.gov/downloads/food/complianceenforcement/sampling/ucm512217.pdf.
3. US Food and Drug Administration, "Microbiological Surveillance Sampling. Testing to Support Prevention Under the FDA Food Safety Modernization Act," December 7, 2018, https://www.fda.gov/Food/ComplianceEnforcement/Sampling/ucm473112.htm#commodities.
4. Christine Haughney, "Why Is The U.S. Government Hassling French Cheesemakers?," *Food Republic*, May 3, 2016, https://www.foodrepublic.com/2016/05/03/why-is-the-u-s-government-waging-a-secret-war-against-french-cheese.
5. Nick Woods, "America's Most Award-Winning Cheese Is Back from the Dead," *Munchies*, January 20, 2016, https://munchies.vice.com/en_us/article/xy7gvn/americas-most-award-winning-cheese-is-back-from-the-dead.
6. Dennis J. D'Amico and Catherine W. Donnelly, "FDA's Domestic and Imported Cheese Compliance Program Results: January 1, 2004–December 31, 2006," *Food Protection Trends* 31, no. 4 (2011): 216–26.
7. Marie Limoges and Catherine W. Donnelly, "FDA's Compliance Program Guideline Criteria for Non-toxigenic *Escherichia coli*: Impacts on Domestic and Imported Cheeses, P2-142," *Journal of Food Protection Supplement* 79 (2016): 191, https://jfoodprotection.org/doi/pdf/10.4315/0362-028X-79.sp1.1.
8. US Food and Drug Administration, "FDA Is Taking a New Look at Criteria for Raw Milk Cheese," *Constituent Update*, February 8, 2016, https://www.fda.gov/Food/NewsEvents/ConstituentUpdates/ucm482438.htm.
9. Warning Letter from Ronald M. Pace, District Director, Public Health Service, U.S. Food and Drug Administration to Finger Lakes

Farmstead Cheese Company LLC (October 23, 2012), http://www.fda.gov/iceci/enforcementactions/warningletters/2012/ucm325714.htm.
10. International Dairy Foods Association, "Geographical Indications," https://www.idfa.org/issues/geographical-indications.
11. International Dairy Foods Association FDA Deadlines and Priorities Top Regulatory RoundUP Agenda March 16, 2016, https://www.idfa.org/news-views/headline-news/article/2016/03/16/fda-deadlines-and-priorities-top-regulatory-roundup-agenda accessed 3/6/2019.
12. Harvard Law Petrie-Flom Center, "FSMA Conference Part 5: International Issues and Trade Implications," March 4, 2014, http://blog.petrieflom.law.harvard.edu/2014/03/04/fsma-conference-part-5-international-issues-and-trade-implications/.
13. Janet Fletcher, "Gone for Good?," *Planet Cheese*, March 10, 2018, http://www.janetfletcher.com/blog/2018/3/10/gone-for-good.
14. Michael R. Taylor and Howard R. Sklamberg, "Internationalizing Food Safety: FDA's Role in the Global Food System," *Harvard International Review* 37, no. 3 (Spring 2016), 32–37, https://www.jstor.org/stable/26445836?read-now=1&seq=2#page_scan_tab_contents.

Chapter 13: Toward a New Regulatory Model

1. Sami L. Gottlieb et al., "Multistate Outbreak of Listeriosis Linked to Turkey Deli Meat and Subsequent Changes in US Regulatory Policy," *Clinical Infectious Diseases* 42, no. 1 (2006): 29–36.
2. US Department of Agriculture Food Safety Inspection Service, "Controlling *Listeria monocytogenes* in Post-Lethality Exposed Ready-To-Eat Meat and Poultry Products" *FSIS Compliance Guideline*, 2014, https://www.fsis.usda.gov/wps/wcm/connect/d3373299-50e6-47d6-a577-e74a1e549fde/Controlling-Lm-RTE-Guideline.pdf?MOD=AJPERES.
3. US Department of Agriculture Food Safety Inspection Service, "Chapter 1, FSIS Listeria Guideline: Requirements of the *Listeria* Rule" (2012), https://www.fsis.usda.gov/wps/wcm/connect/8d49abc3-6a8a-4c9e-a1d3-33befeb5519c/Chapter_1_Controlling_LM_RTE_guideline_0912.pdf?MOD=AJPERES.
4. D. L. Seaman et al., "Modeling the Growth of *Listeria monocytogenes* in Cured Ready-To-Eat Processed Meat Products by Manipulation of

Sodium Chloride, Sodium Diacetate, Potassium Lactate, and Product Moisture Content," *Journal of Food Protection* 65, no. 4 (2002): 651–58.

5. US Department of Agriculture, "*L. monocytogenes* Regulations," November 29, 2016, 38-1, https://www.fsis.usda.gov/wps/wcm/connect/c71a37e9-1fc5-49b8-b7d6-48c67375124f/39_IM_Lm_Regs.pdf?MOD=AJPERES.
6. US Department of Justice, "United States Files Enforcement Action Against Michigan Cheese Company and Owners to Stop Distribution of Adulterated Cheese Products," Office of Public Affairs, August 8, 2014, https://www.justice.gov/opa/pr/united-states-files-enforcement-action-against-michigan-cheese-company-and-owners-stop.
7. US Department of Justice, "United States Files Enforcement Action."
8. *United States of America v. Finger Lakes Farmstead Cheese Company et al.*, Case 1:14-cv-00053-RJA Document 5 Filed 03/20/14, https://www.courtlistener.com/recap/gov.uscourts.nywd.97089.5.0.pdf.
9. Ben Chapman, "Blessed Are the Cheesemakers (Except Those That Have an Import Alert)," September 3, 2014, https://www.barfblog.com/categories/ecoli/page/113/hse.gov.uk/campaigns/%20farmsafe/ecoli/page/134.
10. GMA Update, "Re-introducing the Alliance for Listeriosis Prevention," February 19, 2016, https://www.qualityassurancemag.com/article/re-introducing-the-alliance-for-listeriosis-prevention-february-2016.
11. Produce Marketing Association 2015, "FDA CFSAN Food Advisory Committee; Notice of Meeting Concerning *Listeria monocytogenes*; 80 Fed. Reg. 69229 (9 November 2015), [Docket No. FDA-2015-N-0001]," November 30, 2015, https://www.pma.com/~/media/pma-files/food-safety/final-pma-comments-fda-fac-on-lm-30-nov-2015.pdf?la=en.
12. United States, Executive Office of the President, "Delivering Government Solutions in the 21st Century, Reform Plan and Reorganization Recommendations" (2018), https://www.performance.gov/GovReform/Reform-and-Reorg-Plan-Final.pdf.
13. United States, Executive Office of the President, "Delivering Government Solutions."
14. Safe Food Act of 2015, S. 287, 115th Congress, (2015).
15. Sam P.K. Collins, "President Obama Is Proposing A New Way To Deal With Food Safety. But Will It Work?" *ThinkProgress*, March 6, 2015,

https://thinkprogress.org/president-obama-is-proposing-a-new-way-to-deal-with-food-safety-but-will-it-work-92d484ee609b.

Chapter 14: International Perspectives

1. A. Pouvreau and M. Porter, "French Cheesemakers Crippled by EU Health Measures," *Newsweek*, August 26, 2014, https://www.newsweek.com/2014/08/29/french-cheesemakers-crippled-eu-health-measures-266799.html.
2. Jeanne Carpenter, "Small Cheesemaking Operations Lead Growth in U.S. Cheese Industry" *Cheese Underground*, July 15, 2014, https://cheeseunderground.com/2014/07/15/small-cheesemaking-operations-lead-growth-in-u-s-cheese-industry.
3. Dick Groves, "Judging Confirms Positive Outlook For Specialty Cheese," *Cheese Reporter*, 2017, http://www.cheesereporter.com/editorial2017.htm.
4. Cabot Creamery, "Frequently Asked Questions," https://www.cabotcheese.coop/faq.
5. Dan Strongin, "On Serendipity, Common Sense and Social Justice," *Cheese Reporter*, November 30, 2018, http://www.cheesereporter.com/Strongin/strongin.htm.
6. Strongin, "On Serendipity."
7. Food Standards Agency, "General Food Law," https://www.food.gov.uk/business-guidance/general-food-law#regulation-178-2002-provisions.
8. European Communities, "Regulation (EC) No 852/2004 of the European Parliament and of the Council of 29 April 2004 on the Hygiene of Foodstuffs," 2004, http://eur-lex.europa.eu/legal-content/EN/TXT/?uri=CELEX:02004R0852-20090420.
9. Official Journal of the European Union, "Information and Notices. Commission Notice on the Implementation of Food Safety Management Systems Covering Prerequisite Programs (PRPs) and Procedures Based on the HACCP Principles, Including the Facilitation/Flexibility of the Implementation in Certain Food Businesses," Volume 59, July 30, 2016, http://eur-lex.europa.eu/legal-content/EN/TXT/PDF/?uri=OJ:C:2016:278:FULL&from=EN.
10. Specialist Cheesemakers Association, "Specialist Cheesemakers Association Assured Code of Practice," Edition 1 (2015).

11. Melody Schmid et al., "Risk Factors Involved in the Contamination of Swiss Cheeses by *Listeria monocytogenes* and Coagulase-Positive Staphylococci," *Agroscope Science* 37 (2016): 1–12.
12. Schmid et al., "Risk Factors."

Chapter 15: Advocacy

1. Catherine Donnelly, ed., *Cheese and Microbes* (Washington, D.C.: American Society for Microbiology Press, 2014), 1–333.
2. Corinna Hawkes and Juergen Voegele,"Our Food System Is Broken: Here Are 3 Ways to Fix It," World Economic Forum Annual Meeting, 2018, https://www.weforum.org/agenda/2018/01/our-food-system-is-broken-three-ways-to-fix-it.
3. Thin Lei Win, "France Is Most Food Sustainable Country, U.S. and U.K. Faltering," *Food Quality & Safety*, November 29, 2018, https://www.foodqualityandsafety.com/article/france-is-most-food-sustainable-country-u-s-and-u-k-faltering.
4. Madeleine Vedel, "Saving the Raw Milk Cheeses of Provence," A Campaign for Real Milk, Weston A. Price Foundation, Modified on January 11, 2017, https://www.realmilk.com/international-updates/saving-the-raw-milk-cheeses-of-provence.
5. Bronwen Percival, "Techniques: Raw Milk Microbiology Guide," July 26, 2014, http://microbialfoods.org/techniques-raw-milk-microbiology-preview.
6. F. Driehuis et al., "Silage Review: Animal and Human Health Risks from Silage," *Journal of Dairy Science* 101, no. 5 (2018): 4093–4110.
7. Laura Kiesel, "How Real Cheese Made Its Comeback," *The Atlantic*, June 15, 2016, https://www.theatlantic.com/technology/archive/2016/07/real-cheese-is-back/491402.
8. Slow Cheese, "A Slow Food Campaign," https://www.slowfood.com/slowcheese/welcome_en.lasso.html.
9. Oldways, "Oldways Cheese Coalition," https://oldwayspt.org/programs/oldways-cheese-coalition.

INDEX

Note: Page numbers in italics refer to figures and photographs. Page numbers followed by a "t" refer to tables.

ABOUT THE AUTHOR

Jennifer Francoeur

DR. CATHERINE DONNELLY is a Professor of Nutrition and Food Science at the University of Vermont, and an expert on *Listeria* and other foodborne pathogens. In 2017, Dr. Donnelly won the James Beard Award for Reference and Scholarship for her work as the Editor-in-Chief of *The Oxford Companion to Cheese*, the most comprehensive cheese encyclopedia ever published. Dr. Donnelly is also the editor of the book *Cheese and Microbes* (ASM Press, 2014).